SpringerBriefs in Energy

Computational Modeling of Energy Systems

Series Editors

Thomas Nagel
Haibing Shao

More information about this series at http://www.springer.com/series/8903

Norbert Böttcher • Norihiro Watanabe
Uwe-Jens Görke • Olaf Kolditz

Geoenergy Modeling I

Geothermal Processes in Fractured Porous Media

 Springer

Norbert Böttcher
Helmholtz Centre for Environmental
    Research – UFZ
Leipzig, Germany

Norihiro Watanabe
Helmholtz Centre for Environmental
    Research – UFZ
Leipzig, Germany

Uwe-Jens Görke
Helmholtz Centre for Environmental
    Research – UFZ
Leipzig, Germany

Olaf Kolditz
Helmholtz Centre for Environmental
    Research – UFZ and
    Technische Universität Dresden
Leipzig, Germany

ISSN 2191-5520   ISSN 2191-5539 (electronic)
SpringerBriefs in Energy
ISBN 978-3-319-31333-7   ISBN 978-3-319-31335-1 (eBook)
DOI 10.1007/978-3-319-31335-1

Library of Congress Control Number: 2016935862

Printed on acid-free paper

This Springer imprint is published by Springer Nature
The registered company is Springer International Publishing AG Switzerland

About the Author

Dr. Norbert Böttcher is currently working as a scientist at the Federal Institute for Geosciences and Natural Resources in Hanover, Germany. After his graduation in Water Supply Management and Engineering, Dr. Böttcher had worked at the Institute for Groundwater Management at Dresden University of Technology, Germany, as a lecturer and a research scientist. He received his Ph.D. for developing a mathematical model for the simulation of non-isothermal, compressible multiphase flow processes within the context of subsurface CO_2 storage. As a post-doc, Dr. Böttcher was employed at the Helmholtz-Centre for Environmental Research – UFZ in Leipzig, Germany. His research topics include numerical simulations and model development of geotechnical applications.

Foreword

This tutorial presents the introduction of the open-source software *OpenGeoSys* (OGS) for geothermal applications. The material is based on several national training courses at the Helmholtz Centre of Environmental Research or UFZ in Leipzig, the Technische Universität Dresden and the German Research Centre for Geosciences or GFZ in Potsdam, Germany, but also international training courses on the subject held in South Korea (2012) and China (2013). This tutorial is the result of a close cooperation within the OGS community (www.opengeosys.org). These voluntary contributions are highly acknowledged.

The book contains general information regarding heat transport modelling in porous and fractured media and step-by-step model setup with OGS and related components such as the OGS Data Explorer. Five benchmark examples are presented in detail.

This book is intended primarily for graduate students and applied scientists, who deal with geothermal system analysis. It is also a valuable source of information for professional geoscientists wishing to advance their knowledge in numerical modelling of geothermal processes including thermal convection processes. As such, this book will be a valuable help in the training of geothermal modelling.

There are various commercial software tools available to solve complex scientific questions in geothermics. This book will introduce the user to an open-source numerical software code for geothermal modelling, which can even be adapted and extended based on the needs of the researcher.

This tutorial is the first in a series that will represent further applications of computational modelling in energy sciences. Within this series, the planned tutorials related to the specific simulation platform OGS are as follows:

- OpenGeoSys Tutorial. Basics of Heat Transport Processes in Geothermal Systems, Böttcher et al. (2015), this volume
- OpenGeoSys Tutorial. Shallow Geothermal Systems, Shao et al. (2015[1])

---

[1]Publication time is approximated.

- OpenGeoSys Tutorial. Enhanced Geothermal Systems, Watanabe et al. [2016 (see footnote 1)]
- OpenGeoSys Tutorial. Geotechnical Storage of Energy Carriers, Böttcher et al. [2016 (see footnote 1)]
- OpenGeoSys Tutorial. Models of Thermochemical Heat Storage, Nagel et al. [2017 (see footnote 1)]

These contributions are related to a similar publication series in the field of environmental sciences, namely:

- Computational Hydrology I: Groundwater flow modeling, Sachse et al. (2015), DOI 10.1007/978-3-319-13335-5, http://www.springer.com/de/book/9783319133348,
- OpenGeoSys Tutorial. Computational Hydrology II: Density-dependent flow and transport processes, Walther et al. [2016 (see footnote 1)],
- OGS Data Explorer, Rink et al. [2016 (see footnote 1)],
- Reactive Transport Modeling I [2017 (see footnote 1)],
- Multiphase Flow [2017 (see footnote 1)].

Leipzig, Germany   Olaf Kolditz
Leipzig, Germany   Norbert Böttcher
Leipzig, Germany   Uwe-Jens Görke
Leipzig, Germany   Norihiro Watanabe
June 2015

Acknowledgements

We deeply acknowledge the continuous scientific and financial support to the *OpenGeoSys* development activities by the following institutions:

We would like to express our sincere thanks to HIGRADE in providing funding for the *OpenGeoSys* training course at the Helmholtz Centre for Environmental Research.

We also wish to thank the *OpenGeoSys*-developer group (ogs-devs@google groups.com) and the users (ogs-users@googlegroups.com) for their technical support.

Contents

Chapter 1
Geothermal Energy

Welcome to the OGS HIGRADE Tutorials on Computational Energy Systems. The first tutorial will introduce the reader to the field of modeling geothermal energy systems. In the beginning chapter we will introduce geothermal systems, their utilization, geothermal processes as well as the open source simulation software OpenGeoSys (OGS @ www.opengeosys.org).

1.1 Geothermal Systems

"Geothermal energy is a promising alternative energy source as it is suited for base-load energy supply, can replace fossil fuel power generation, can be combined with other renewable energy sources such as solar thermal energy, and can stimulate the regional economy" is cited from the Editorial to a new open access journal Geothermal Energy (Kolditz et al. 2013) in order to appraise the potential of this renewable energy resource for both heat supply and electricity production.

Geothermal energy became an essential part in many research programmes world-wide. The current status of research on geoenergy (including both geological energy resources and concepts for energy waste deposition) in Germany and other countries recently was compiled in a thematic issue on "Geoenergy: new concepts for utilization of geo-reservoirs as potential energy sources" (Scheck-Wenderoth et al. 2013). The Helmholtz Association dedicated a topic on geothermal energy systems into its next five-year-program from 2015 to 2019 (Huenges et al. 2013).

Looking at different types of geothermal systems it can be distinguished between shallow, medium, and deep systems in general (Fig. 1.1). Installations of shallow systems are allowed down to 100 m by regulation, and include soil and shallow aquifers, therefore. Medium systems are associated with hydrothermal resources and may be suited for underground thermal storage (Bauer et al. 2013). Deep systems are connected to petrothermal sources and need to be stimulated to

© Springer International Publishing Switzerland 2016
N. Böttcher et al., *Geoenergy Modeling I*, SpringerBriefs in Energy,
DOI 10.1007/978-3-319-31335-1_1

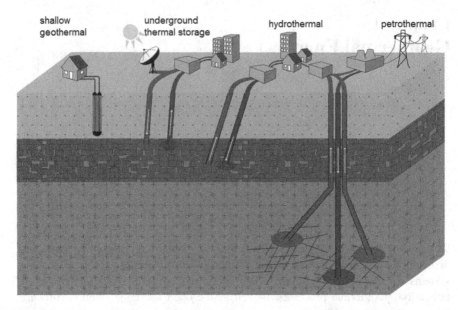

Fig. 1.1 Overview of different types of geothermal systems: shallow, mid and deep systems (Huenges et al. 2013)

increase hydraulic conductivity for heat extraction by fluid circulation (Enhanced Geothermal Systems—EGS). In general, the corresponding temperature regimes at different depths depend on the geothermal gradient (Clauser 1999). Some areas benefit from favourable geothermal conditions with amplified heat fluxes, e.g., the North German Basin, Upper Rhine Valley and Molasse Basin in Germany (Cacace et al. 2013).

1.2 Geothermal Resources

Conventional geothermal systems mainly rely on near-surface heated water (hydrothermal systems) and are regionally limited to near continental plate boundaries and volcanos. Figure 1.2 shows the Earth pattern of plates, ridges, subduction zones as well as geothermal fields.

Figure 1.3 shows an overview map of hydrothermal systems in China including a classification to high, mid and low-temperature reservoirs and basins (Kong et al. 2014). Current research efforts concerning hydrothermal resources focus on the sustainable development of large-scale geothermal fields. Pang et al. (2012) designed a roadmap of geothermal energy development in China and reported the recent progress in geothermal research in China. Recently, Pang et al. (2015) presented a new classification of geothermal resources based on the type of heat source and followed by the mechanisms of heat transfer. A new Thematic Issue of

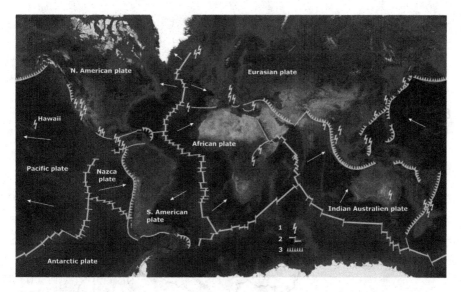

Fig. 1.2 World pattern of plates, oceanic ridges, oceanic trenches, subduction zones, and geothermal fields. *Arrows* show the direction of movement of the plates towards the subduction zones. (1) Geothermal fields producing electricity; (2) mid-oceanic ridges crossed by transform faults (long transversal fractures); (3) subduction zones, where the subducting plate bends downwards and melts in the asthenosphere (Dickson and Fanelli 2004)

the Environmental Earth Sciences journal is focusing on petrothermal resources and particularly enhanced geothermal systems (Kolditz et al. 2015b).

Germany's geothermal resources are mainly located within the North German Basin, the upper Rhine Valley (border to France) and the Bavarian Molasses. Figure 1.4 also depicts the existing geothermal power plants as well as geothermal research and exploration sites.

1.3 Geothermal Processes

Where does geothermal energy come from? Per definition "Geothermal Energy" refers to the heat stored within the solid Earth (Fig. 1.5). The heat sources of geothermal energy are:

- From residual heat from planetary accretion (20 %)
- From radioactive decay (80 %)

Some interesting numbers about geothermal temperatures are:

- The mean surface temperature is about 15 °C.
- The temperature in the Earth's centre is 6000–7000 °C hot (99 % of the Earth's volume is hotter than 1000 °C)
- The geothermal gradient in the upper part is about 30 K per kilometer depth.

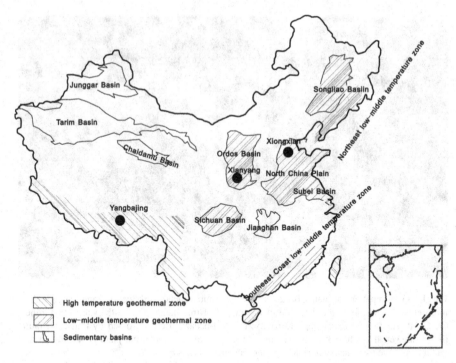

Fig. 1.3 Hydrothermal system in China—a classification by high and mid-low temperature reservoirs and basis (Kong et al. 2014)

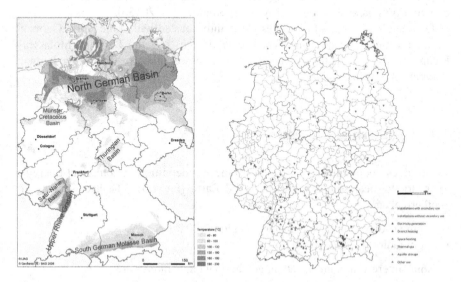

Fig. 1.4 Regions with hydrothermal resources and geothermal installations in Germany (Agemar et al. 2014)

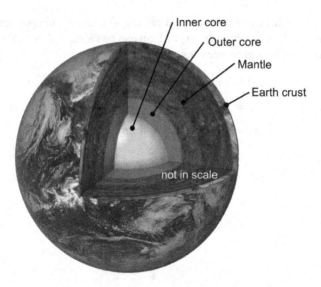

Fig. 1.5 Structure of earth

An Exercise Before We Start ...

The (total) energy content of the Earth is 10^{31} J. How can we estimate such a number?

$$Q = cmT \tag{1.1}$$

with

- Q amount of heat in J,
- heat capacity c in $J\,kg^{-1}\,K^{-1}$,
- mass m in kg and
- temperature T in K.

The amount of heat released or spent related to a certain temperature (100 K) change from a rock cube with side length 1 km can be calculated as follows (we use typical number for rock properties, e.g. granite):

$$\Delta Q = cm\Delta T \tag{1.2}$$

$$c = 790\,\mathrm{J\,kg^{-1}\,K^{-1}}$$

$$m = \rho V = 2.65 \cdot 10^3\,\mathrm{kg\,m^{-3}} \cdot 10^9\,\mathrm{m^3}$$

$$= 2.65 \cdot 10^{12}\,\mathrm{kg}$$

$$\Delta Q = 2.0935 \cdot 10^{17}\,\mathrm{J}$$

$$= 2.0935 \cdot 10^8\,\mathrm{GJ} \tag{1.3}$$

The "heat power production" from deep geothermal energy resources in the USA in 2010 was about 2000 MW. The amount of heat or heat energy (in Joule) gained from a certain heat power (in Watt) is defined as the product of heat power by time (see calculation below.) We would like to find out the volume of the rock cube corresponding to the annual geothermal energy production of the USA in 2010 in order to get a better impression about the potential of deep geothermal systems.

$$[\mathrm{J}] = [\mathrm{W\,s}] \tag{1.4}$$

$$Q = 2 \cdot 10^9\,\mathrm{W} \cdot 365\,\mathrm{d} \cdot 86400\,\mathrm{s\,d^{-1}}$$

$$Q = 6.3072 \cdot 10^{10}\,\mathrm{MJ}$$

$$Q_{\mathrm{US}} = 0.30127\,\mathrm{J_{km}}$$

$$Q_{\mathrm{US}} = \mathrm{J_{670.4\,m}} \tag{1.5}$$

The result is that the annual geothermal energy production of the USA in 2010 corresponds to heat released from a rock cube with length of 670 m (the temperature change was assumed to be $\Delta T = 100\,\mathrm{K}$). This is approximately the length of the UFZ campus.

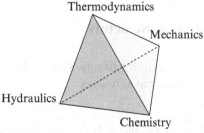

Fig. 1.6 THMC coupling concept

1.4 OpenGeoSys (OGS)

OGS is a scientific open-source initiative for numerical simulation of thermo-hydro-mechanical/chemical (THMC) processes in porous and fractured media, continuously developed since the mid-eighties. The OGS code is targeting primarily applications in environmental geoscience, e.g. in the fields of contaminant hydrology, water resources management, waste deposits, or geothermal systems, but it has also been applied to new topics in energy storage recently (Fig. 1.6).

OGS is participating several international benchmarking initiatives, e.g. DEVO-VALEX (with applications mainly in waste repositories), CO2BENCH (CO2 storage and sequestration), SeSBENCH (reactive transport processes) and HM-Intercomp (coupled hydrosystems)

The basic concept is to provide a flexible numerical framework (using primarily the Finite Element Method (FEM)) for solving coupled multi-field problems in porous-fractured media. The software is written with an object-oriented (C++) FEM concept including a broad spectrum of interfaces for pre- and postprocessing. To ensure code quality and to facilitate communications among different developers worldwide OGS is outfitted with professional software-engineering tools such as platform-independent compiling and automated result testing tools. A large benchmark suite has been developed for source code and algorithm verification over the time. Heterogeneous or porous-fractured media can be handled by dual continua or discrete approaches, i.e. by coupling elements of different dimensions. OGS has a built-in random-walk particle tracking method for Euler-Lagrange simulations. The code has been optimized for massive parallel machines. The OGS Tool-Box concept promotes (mainly) open source code coupling e.g. to geochemical and biogeochemical codes such as iPHREEQC, GEMS, and BRNS for open functionality extension. OGS also provides continuous workflows including various interfaces for pre- and post-processing. Visual data integration became an important tool for establishing and validating data driven models (OGS DataExplorer). The OGS software suite provides three basic modules for data integration, numerical simulation and 3D visualization.

1.5 Tutorial and Course Structure

The three-day course "Computational Energy Systems I: Geothermal Processes" contains 11 units and it is organized as follows:

- **Day 1:**
  - OGS-CES-I-01: Lecture: Geothermal energy systems (Sect. 1.1–1.3)
  - OGS-CES-I-02: Exercise: Geothermal energy systems (Sect. 1.4)
  - OGS-CES-I-03: Lecture: Theory of heat transport processes in porous media (Chap. 2)
- **Day 2:**
  - OGS-CES-I-04: Lecture: Introduction to numerical methods (Sect. 3.1–3.4)
  - OGS-CES-I-05: Lecture: Finite element method (Sect. 3.5)
  - OGS-CES-I-06: Exercise: Heat conduction in a semi-finite domain (Sect. 4.3)
  - OGS-CES-I-07: Exercise: Heat flux through a layered porous medium (Sect. 4.4)
- **Day 3:**
  - OGS-CES-I-08: Exercise: Heat transport in a porous medium (Sect. 4.5)
  - OGS-CES-I-09: Exercise: Heat transport in a porous-fractured medium (Sect. 4.6)
  - OGS-CES-I-10: Exercise: Heat convection in a porous medium (Sect. 4.7)
  - OGS-CES-I-11: Lecture: Introduction to geothermal case studies

The material is based on:

- OGS training course on geoenergy aspects held by Norihiro Watanabe in November 2013 in Guangzhou
- OGS training course on CO_2-reduction modelling held by Norbert Böttcher in 2012 in Daejon, South Korea
- OGS benchmarking chapter on heat transport processes by Norbert Böttcher
- University lecture material (TU Dresden) and presentations by Olaf Kolditz

Chapter 2
Theory

In this chapter we briefly glance at basic concepts of porous medium theory (Sect. 2.1.1) and thermal processes of multiphase media (Sect. 2.1.2). We will study the mathematical description of thermal processes in the context of continuum mechanics and numerical methods for solving the underlying governing equations (Sect. 2.2).

2.1 Continuum Mechanics of Porous Media

There is a great body of existing literature in the field of continuum mechanics and thermodynamics of porous media existing, you should look at, e.g. Carslaw and Jaeger (1959), Häfner et al. (1992), Bear (1972), Prevost (1980), de Boer (2000), Ehlers and Bluhm (2002), Kolditz (2002), Lewis et al. (2004), Goerke et al. (2012), Diersch (2014).

2.1.1 Porous Medium Model

"The Theory of Mixtures as one of the basic approaches to model the complex behavior of porous media has been developed over decades (concerning basic assumptions see e.g. Bowen 1976; Truesdell and Toupin 1960). As the Theory of Mixtures does not incorporate any information about the microscopic structure of the material,[1] it has been combined with the Concept of Volume Fractions by e.g.

---

[1] Within the context of the Theory of Mixtures the ideal mixture of all constituents of a multiphase medium is postulated. Consequently, the realistic modeling of the mutual interactions of the constituents is difficult.

© Springer International Publishing Switzerland 2016

N. Böttcher et al., *Geoenergy Modeling I*, SpringerBriefs in Energy,

DOI 10.1007/978-3-319-31335-1_2

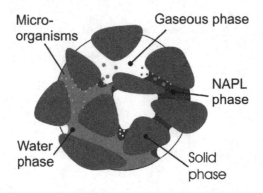

Bowen (1980), de Boer and Ehlers (1986), Lewis and Schrefler (1998), Prevost (1980). Within the context of this enhanced Theory of Mixtures (also known as Theory of Porous Media), all kinematical and physical quantities can be considered at the macroscale as local statistical averages of their values at the underlying microscale. Concerning a detailed overview of the history of the modeling of the behavior of multiphase multicomponent porous media, the reader is referred to e.g. de Boer (2000). Comprehensive studies about the theoretical foundation and numerical algorithms for the simulation of coupled problems of multiphase continua are given in e.g. de Boer (2000), Ehlers and Bluhm (2002), Lewis and Schrefler (1998) and the quotations therein." (Kolditz et al. 2012)

- A porous medium consists of different phases, i.e. a solid and at least one fluid phase,
- heat transfer processes are diffusion (solid phase), advection, dispersion,
- heat transfer between solid grains can occur by radiation when the fluid is a gas.
- A valid averaging volume for a porous medium is denoted as a representative elementary volume (REV).

Since the geometry of porous media in reality is not known exactly, a continuum approach comes into play. Figure 2.1 depicts the general idea of the porous medium approach. We do not need to know all the details about the microscopic porous medium structure but the portions of each phase which can be described macroscopically by **porosity** and **saturation**.

2.1.2 Thermal Processes

The assumption of local thermodynamic equilibrium is important and valid for many geothermal applications. At low Reynolds number flows and with small grain diameters, we may neglect the difference in temperature between the individual

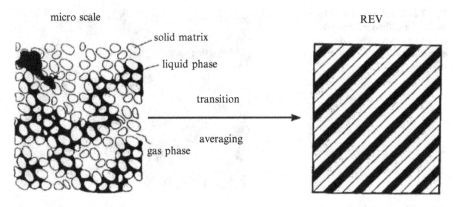

Fig. 2.1 REV concept (Helmig 1997)

phases, i.e. all phase temperatures are assumed to be equal. Physically this means, that the energy exchange between the phases is significantly faster than the energy transport within a phase.

The most important thermal processes within the context of geothermal energy systems are heat diffusion, heat advection, and heat storage. Additionally we might have to consider:

- heat radiation
- latent heat
- heat sources and sinks

2.1.2.1   Heat Diffusion

Diffusion processes basically are resulting from the Brownian molecular motion.

Heat diffusion basically is the transportation of heat by molecular activity. The diffusive heat flux is described by the famous Fouriers[2] law (2.1) (Fig. 2.2).

$$\mathbf{j}_{\text{diff}} = -\lambda^{\text{eff}} \nabla T \tag{2.1}$$

where $\mathbf{j}_{\text{diff}}$ is the conductive heat flux, and λ^{eff} is the effective thermal conductivity of a porous media. The simplest definition of a porous medium property is given by volume fraction weighting. In case of a fully water saturated porous medium with porosity n, λ^{eff} reads

$$\lambda^{\text{eff}} = n\lambda^{\text{w}} + (1-n)\lambda^{\text{s}} \tag{2.2}$$

---

[2]A bit history of Fourier comes in the lecture.

Fig. 2.2 Transport of heat by molecular diffusion. (Picture from http://www.ehow.de/experimente-konduktion-strategie_9254/)

Table 2.1 Typical values of thermal conductivity in [W m^{-1} K^{-1}] at 20 °C for some natural media

Material	Value
Iron	73
Limestone	1.1
Water	0.58
Dry air	0.025

where superscripts ^{s} and ^{w} stand for solid phase and water phase, respectively. For unsaturated media, a general formulation for effective heat conductivity can be written as

$$\lambda^{\text{eff}} = n \sum_{f} S^f \lambda^f + (1 - n) \lambda^s \tag{2.3}$$

where the superscript f stands for any fluid phase residing in the medium. A few typical examples of thermal conducivity for some natural media is listed in Table 2.1.

2.1.2.2 Heat Advection

Heat advection is the transportation of heat by fluid motion.[3]

The advective heat flux is described as

$$\mathbf{j}_{\text{adv}} = c^w \rho^w T \mathbf{v} \tag{2.4}$$

where c^w is specific heat and ρ^w is density of the water phase, and $\mathbf{v}$ is the Darcy velocity (discharge per unit area) (Fig. 2.3).

---

[3]Note: "Heat" can be even cold in case the temperature is lower than the ambient one ;-).

Fig. 2.3 Transport of heat by fluid motion. (Picture from http://notutopiacom.ipage. com/wordpress/about-2/ about/)

2.1.2.3 Heat Storage

Heat storage in a porous medium can be expressed by the amount of heat Q in [J] within a balance volume V in [m^3]

$$Q = (c\rho)^{\text{eff}} VT \tag{2.5}$$

where c is the specific heat capacity of the medium with

$$c_p = \left.\frac{\partial h}{\partial T}\right|_{p=\text{const}} \quad c_v = \left.\frac{\partial u}{\partial T}\right|_{v=\text{const}} \tag{2.6}$$

where c_p and c_v are isobaric and isochoric heat capacities, respectively. The specific heat capacity c is the heat necessary to increase the temperature of a unit mass of a medium by 1 K. The heat capacity of a porous medium is composed by its phase properties and can be expressed as an effective parameter, obtained analogously to (2.2):

$$(c\rho)^{\text{eff}} = nc^{\text{w}}\rho^{\text{w}} + (1-n)c^{\text{s}}\rho^{\text{s}} \tag{2.7}$$

2.1.2.4 Heat Dispersion

Similar to mass dispersion, the structure of a porous medium results in a dispersive transport of heat. The basic idea behind the classic hydrodynamic dispersion theory by Scheidegger (Scheidegger 1961) is a normal distribution pattern through a regular porous medium (Fig. 2.4).

$$\mathbf{j}_{\text{disp}} = -\rho^{\text{w}}c^{\text{w}}\left(\alpha_{\text{T}}n|\mathbf{v}|\delta_{ij} + (\alpha_{\text{L}} - \alpha_{\text{T}})n\frac{v_i v_j}{|\mathbf{v}|}\right)\nabla T \tag{2.8}$$

where α_{T} and α_{L} are transversal and longitudinal dispersivity.

Fig. 2.4 Source: Naumov,
personal communication

2.1.2.5   Heat Balance

Flow rates of heat (energy) in porous media can be described by balance equations
The heat balance equation is expressing an equilibrium of thermal processes, i.e.
heat storage, diffusive and advective fluxes as well as heat sources and sinks.

$$c\rho\frac{\partial T}{\partial t} + \nabla \cdot (\mathbf{j}_{\text{diff}} + \mathbf{j}_{\text{adv}} + \mathbf{j}_{\text{disp}}) = Q_{\text{T}} \qquad (2.9)$$

where Q_{T} is the heat production term in $[\text{J}\,\text{m}^{-3}\,\text{s}^{-1}]$.

2.2   Governing Equations of Heat Transport in Porous Media

2.2.1   Energy Balance

The equation of energy conservation is derived from the first law of thermodynamics
which states that the variation of the total energy of a system is due to the work of
acting forces and heat transmitted to the system.

The total energy per unit mass e (specific energy) can be defined as the sum of
internal (thermal) energy u and specific kinetic energy $\mathbf{v}^2/2$. Internal energy is due
to molecular movement. Gravitation is considered as an energy source term, i.e. a
body force which does work on the fluid element as it moves through the gravity
field. The conservation quantity for energy balance is total energy density

$$\psi^e = \rho e = \rho(u + \mathbf{v}^2/2) \qquad (2.10)$$

Using mass and momentum conservation we can derive the following balance equation for the internal energy

$$\rho \frac{du}{dt} = \rho q'' - \nabla \cdot (\mathbf{j}_{\text{diff}} + \mathbf{j}_{\text{disp}}) + \boldsymbol{\sigma} : \nabla \mathbf{v} \tag{2.11}$$

where q'' is the internal energy (heat) source, $\mathbf{j}_{\text{diff}}$ and $\mathbf{j}_{\text{disp}}$ are the diffusive and dispersive heat fluxes, respectively. Utilizing the definition of the material derivative

$$\frac{dT}{dt} = \frac{\partial T}{\partial t} + \mathbf{v} \cdot \nabla T \tag{2.12}$$

and neglecting stress power, we obtain the heat energy balance equation for an arbitrary phase

$$\rho c \frac{\partial T}{\partial t} + \rho c \mathbf{v} \cdot \nabla T - \nabla \cdot \lambda \nabla T = \rho q_{\text{T}} \tag{2.13}$$

where λ contains both the diffusive and dispersive heat conduction parts.

2.2.2 Porous Medium

The heat balance equation for the porous medium consisting of several solid and fluid phases is given by

$$\sum_{\alpha} \epsilon^{\alpha} c^{\alpha} \rho^{\alpha} \frac{\partial T}{\partial t} + \nabla \cdot \left(\sum_{\gamma} n S^{\gamma} \rho^{\gamma} c^{\gamma} \mathbf{v}^{\gamma} \, T - \sum_{\alpha} \epsilon^{\alpha} \lambda^{\alpha} \, \nabla T \right) =$$

$$\sum_{\alpha} \epsilon^{\alpha} \rho^{\alpha} \, q_{\text{th}} \tag{2.14}$$

where α is all phases and γ is fluid phases, and ϵ^{α} is the volume fraction of the phase α.

Most important is the assumption of local thermodynamic equilibrium, meaning that all phase temperatures are equal and, therefore, phase contributions can be superposed. The phase change terms are canceled out with the addition of the individual phases.

With the following assumptions:

- local thermal equilibrium,
- fully saturated porous medium,
- neglecting viscous dissipation effects,

the governing equations for heat transport in a porous medium can be further simplified.

$$(c\rho)^{\text{eff}} \frac{\partial T}{\partial t} + (c\rho)^{\text{fluid}} \mathbf{v} \cdot \nabla T - \nabla \cdot (\lambda^{\text{eff}} \nabla T) = q_{\text{T}} \tag{2.15}$$

with

$$(c\rho)^{\text{eff}} = \sum_{\alpha} \epsilon^{\alpha} c^{\alpha} \rho^{\alpha} \tag{2.16}$$

$$(c\rho)^{\text{fluid}} = n \sum_{\gamma} S^{\gamma} c^{\gamma} \rho^{\gamma} \tag{2.17}$$

$$\lambda^{\text{eff}} = \sum_{\alpha} \epsilon^{\alpha} \lambda^{\alpha} \tag{2.18}$$

For isotropic heat conduction without heat sources and we have the following classic diffusion equation

$$\frac{\partial T}{\partial t} = \nabla \cdot (\alpha^{\text{eff}} \nabla T) \tag{2.19}$$

with heat diffusivity $\alpha^{\text{eff}} = \lambda^{\text{eff}} / (c\rho)^{\text{eff}}$

2.2.2.1   Boundary Conditions

In order to specify the solution for the heat balance equation (2.9) we need to prescribe boundary conditions along all boundaries. Normally we have to consider three types of boundary conditions:

1. Prescribed temperatures (Dirichlet condition)

$$T = \bar{T} \quad \text{on} \quad \Gamma_{\text{T}} \tag{2.20}$$

2. Prescribed heat fluxes (Neumann condition)

$$q_{\text{n}} = \mathbf{j}_{\text{diff}} \cdot \mathbf{n} \quad \text{on} \quad \Gamma_{\text{q}} \tag{2.21}$$

3. Convective heat transfer (Robin condition)

$$q_{\text{n}} = a(T - T_{\infty}) \quad \text{on} \quad \Gamma_{\text{a}} \tag{2.22}$$

where a is the heat transfer coefficient in $[\text{W K}^{-1} \text{ m}^{-2}]$.

2.2.3   Darcy's Law

For linear momentum conservation in porous media with a rigid solid phase we assume, in general, that inertial forces can be neglected (i.e. $d\mathbf{v}/dt \approx 0$) and body forces are gravity at all. Assuming furthermore that internal fluid friction is small in comparison to friction on the fluid-solid interface and that turbulence effects can be neglected we obtain the Darcy law for each fluid phase γ in multiphase flow.

$$\mathbf{q}^\gamma = nS^\gamma \mathbf{v}^\gamma = -nS^\gamma \left(\frac{k_{rel}^\gamma \mathbf{k}}{\mu^\gamma} (\nabla p^\gamma - \rho^\gamma \mathbf{g}) \right) \tag{2.23}$$

Chapter 3
Numerical Methods

3.1 Approximation Methods

There are many alternative methods to solve initial-boundary-value problems arising from flow and transport processes in subsurface systems. In general these methods can be classified into analytical and numerical ones. Analytical solutions can be obtained for a number of problems involving linear or quasi-linear equations and calculation domains of simple geometry. For non-linear equations or problems with complex geometry or boundary conditions, exact solutions usually do not exist, and approximate solutions must be obtained. For such problems the use of numerical methods is advantageous. In this chapter we use the Finite Difference Method to approximate time derivatives. The Finite Element Method as well as the Finite Volume Method are employed for spatial discretization of the region. The Galerkin weighted residual approach is used to provide a weak formulation of the PDEs. This methodology is more general in application than variational methods. The Galerkin approach works also for problems which cannot be casted in variational form.

Figure 3.1 shows an overview on approximation methods to solve partial differential equations together with the associated boundary and initial conditions. There are many alternative methods for solving boundary and initial value problems. In general, these method can be classified as discrete (numerical) and analytical ones.

3.2 Solution Procedure

For a specified mechanical problem the governing equations as well as initial and boundary conditions will be known. Numerical methods are used to obtain an approximate solution of the governing equations with the corresponding initial and boundary conditions. The procedure of obtaining the approximate solution consists

© Springer International Publishing Switzerland 2016
N. Böttcher et al., *Geoenergy Modeling I*, SpringerBriefs in Energy,
DOI 10.1007/978-3-319-31335-1_3

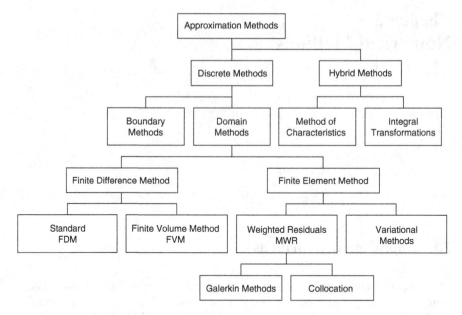

Fig. 3.1 Overview of approximation methods and related sections for discussion (Kolditz 2002)

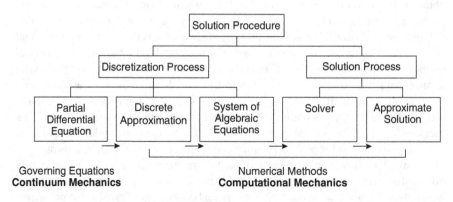

Fig. 3.2 Steps of the overall solution procedure (Kolditz 2002)

of two steps that are shown schematically in Fig. 3.2. The first step converts the continuous partial differential equations and auxiliary conditions (IC and BC) into a discrete system of algebraic equations. This first step is called discretization. The process of discretization is easily identified if the finite difference method is used but it is slightly less obvious with more complicated methods as the finite element method (FEM), the finite volume method (FVM), and combined Lagrangian-Eulerian methods (method of characteristics, operator split methods).

The replacement of partial differential equations (PDE) by algebraic expressions introduces a defined truncation error. Of course it is of great interest to chose algebraic expressions in a way that only small errors occur to obtain accuracy. Equally important as the error in representing the differentiated terms in the governing equation is the error in the solution. Those errors can be examined as shown in Sect. 3.3.

The second step of the solution procedure, shown in Fig. 3.2, requires the solution of the resulting algebraic equations. This process can also introduce an error but this is usually small compared with those involved in the above mentioned discretization step, unless the solver scheme is unstable. The approximate solution of the PDE is exact solution of the corresponding system of algebraic equations (SAE).

3.3 Theory of Discrete Approximation

3.3.1 Terminology

In the first part of this script we developed the governing equations for fluid flow, heat and mass transfer from basic conservation principles. We have seen that hydromechanical field problems (as well as mechanical equilibrium problems) have to be described by partial differential equations (PDEs). The process of translating the PDEs to systems of algebraic equations is called—discretization (Fig. 3.3). This discretization process is necessary to convert PDEs into an equivalent system of algebraic equations that can be solved using computational methods.

$$L(u) = \hat{L}(\hat{u}) = 0 \qquad (3.1)$$

In the following, we have to deal with discrete equations $\hat{L}$ and with discrete solutions $\hat{u}$.

An important question concerning the overall solution procedure for discrete methods is what guarantee can be given that the approximate solution will be close to the exact one of the PDE. From truncation error analysis, it is expected that more accurate solutions could be obtained on refined grids. The approximate solution should converge to the exact one as time step sizes and grid spacing shrink to zero. However, convergence is very difficult to obtain directly, so that usually two steps are required to achieve convergence:

$$\boxed{\text{Consistency} + \text{Stability} = \text{Convergence}}$$

This formula is known as the **Lax equivalence axiom**. That means, the system of algebraic equations resulting from the discretization process should be consistent with the corresponding PDE. Consistency guarantees that the PDE is represented by the algebraic equations. Additionally, the solution process, i.e. solving the system of algebraic equations, must be stable.

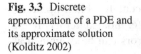

Fig. 3.3 Discrete
approximation of a PDE and
its approximate solution
(Kolditz 2002)

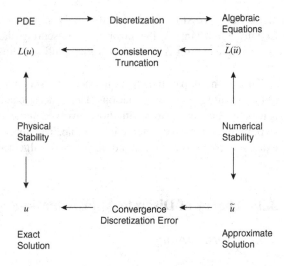

Figure 3.3 presents a graphic to illustrate the relationship between the above introduced basic terms of discrete approximation theory: convergence, stability, truncation, and consistency. These fundamental terms of discrete mathematics are explained further and illustrated by examples in the following.

3.3.2   Errors and Accuracy

The following discussion of convergence, consistency, and stability is concerned with the behavior of the approximate solution if discretization sizes (Δt, Δx) tends to zero. In practice, approximate solutions have to be obtained on finite grids which must be calculable on available computers. Therefore, errors and achievable accuracy are of great interest.

If we want to represent continuous systems with the help of discrete methods, of course, we introduce a number of errors. Following types of errors may occur: solution error, discretization error, truncation error, and round-off errors. Round-off errors may result from solving equation systems. Truncation errors are omitted from finite difference approximations. This means, the representation of differentiated terms by algebraic expressions connecting nodal values on a finite grid introduces a certain error. It is desirable to choose the algebraic terms in a way that only errors as small as possible are invoked. The accuracy of replaced differentiated terms by algebraic ones can be evaluated by considering the so-called truncation error. Truncation error analysis can be conducted by Taylor series expansion (TSE). However, the evaluation of this terms in the TSE relies on the exact solution being known. The truncation error is likely to be a progressively more accurate indicator of the solution error as the discretization grid is refined.

There exist two techniques to evaluate accuracy of numerical schemes. At first, the algorithm can be applied to a related but simplified problem, which possesses an analytical solution (e.g. Burgers equation which models convective and diffusive momentum transport). The second method is to obtain solutions on progressively refined grids and to proof convergence. In general, accuracy can be improved by use of higher-order schemes or grid refinement.

3.3.3 Convergence

Definition: A solution of the algebraic equations which approximate a given PDE is said to be convergent if the approximate solution approaches the exact solution of the PDE for each value of the independent variable as the grid spacing tends to zero. Thus we require

$$\lim_{\Delta t, \Delta x \to 0} | u_j^n - u(t_n, x_j) | = 0 \tag{3.2}$$

Or in other words, the approximate solution converges to the exact one as the grid sizes becomes infinitely small. The difference between exact and approximate solution is the solution error, denoted by

$$\varepsilon_j^n = | u_j^n - u(t_n, x_j) | \tag{3.3}$$

The magnitude of the solution error typically depends on grid spacing and approximations to the derivatives in the PDE.

Theoretical proof of convergence is generally difficult, except very simple cases. For indication of convergence, comparison of approximate solutions on progressively refined grids is used in practice. For PDEs which possesses an analytical solution, like the 1-D advection-diffusion problem, it is possible to test convergence by comparison of numerical solutions on progressively refine grids with the exact solution of the PDE.

3.3.4 Consistency

Definition: The system of algebraic equations (SAE) generated by the discretization process is said to be consistent with the original partial differential equation (PDE) if, in the limit that the grid spacing tends to zero, the SAE is equivalent to the PDE at each grid point. Thus we require

$$\lim_{\Delta t, \Delta x \to 0} | \hat{L}(u_j^n) - L(u[t_n, x_j]) | = 0 \tag{3.4}$$

Or in other words, the SAE converges to the PDE as the grid size becomes zero. Obviously, consistency is necessary for convergence of the approximate solution. However, consistency is not sufficient to guarantee convergence. Although the SAE might be consistent, it does not follow that the approximate solution converges to the exact one, e.g. for unstable schemes. As an example, solutions of the FTCS algorithm diverge rapidly if the scheme is weighted backwards ($\theta > 0.5$). This example emphases that, as indicated by the Lax-Equivalence-Axiom, both consistency and stability are necessary for convergence. Consistency analysis can be conducted by substitution of the exact solution into the algebraic equations resulting from the discretization process. The exact solution is represented as a TSE. Finally, we obtain an equation which consists the original PDE plus a reminder. For consistency the reminder should vanish as the grid size tends to zero.

3.3.5 Stability

Frequently, the matrix method and the von Neumann method are used for stability analysis. In both cases possible growth of the error between approximate and exact solution will be examined. It is generally accepted that the matrix method is less reliable as the von Neumann method. Using the von Neumann method, error at one time level is expanded as a finite Fourier series. For this purpose, initial conditions are represented by a Fourier series. Each mode of the series will grow or decay depending on the discretization. If a particular mode grows without bounds, then an unstable solution exists for this discretization.

3.4 Solution Process

We recall, that the overall solution procedure for PDEs consists of the two major steps: discretization and solution processes (Fig. 3.2). In this section we give a brief introduction to the solution process for equation systems, resulting from discretization methods such as finite difference (FDM), finite element (FEM) and finite volume methods (FVM) (see Chaps. 6–8 in Kolditz 2002). More details on the solution of equation systems can be found e.g. in Hackbusch (1991), Schwetlick and Kretzschmar (1991), Knabner and Angermann (2000), Wriggers (2001).

Several problems in environmental fluid dynamics lead to non-linear PDEs such as non-linear flow (Chap. 12), density-dependent flow (Chap. 14), multi-phase flow (Chaps. 15, 16) in Kolditz (2002). The resulting algebraic equation system can be written in a general way, indicating the dependency of system matrix $\mathbf{A}$ and right-hand-side vector $\mathbf{b}$ on the solution vector $\mathbf{x}$. Consequently, it is necessary to employ iterative methods to obtain a solution.

$$\mathbf{A}(\mathbf{x})\,\mathbf{x} - \mathbf{b}(\mathbf{x}) = \mathbf{0} \qquad (3.5)$$

In the following we consider methods for solving linear equation systems (Sect. 3.4.1) and non-linear equation systems (Sect. 3.4.2).

3.4.1 Linear Solver

The linear version of Eq. (3.5) is given by

$$\mathbf{A}\mathbf{x} - \mathbf{b} = \mathbf{0} \tag{3.6}$$

In general there are two types of methods: direct and iterative algorithms. Direct methods may be advantageous for some non-linear problems. A solution will be produced even for systems with ill-conditioned matrices. On the other hand, direct schemes are very memory consuming. The required memory is in the order of $O(nb^2)$, with n the number of unknowns and b the bandwidth of the system matrix. Therefore, it is always useful to apply algorithms for bandwidth reduction. Iterative methods have certain advantages in particular for large systems with sparse matrices. They can be very efficient in combination with non-linear solver.

The following list reveals an overview on existing methods for solving linear algebraic equation systems.

- Direct methods

  - Gaussian elimination
  - Block elimination (to reduce memory requirements for large problems)
  - Cholesky decomposition
  - Frontal solver

- Iterative methods

  - Linear steady methods (Jacobian, Gauss-Seidel, Richardson and block iteration methods)
  - Gradient methods (CG) (also denoted as Krylov subspace methods)

3.4.1.1 Direct Methods

Application of direct methods to determine the solution of Eq. (3.6)

$$\mathbf{x} = \mathbf{A}^{-1}\mathbf{b} \tag{3.7}$$

requires an efficient techniques to invert the system matrix.

As a first example we consider the Gaussian elimination technique. If matrix $\mathbf{A}$ is not singular (i.e. $\det \mathbf{A} \neq 0$), can be composed in following way.

$$\mathbf{P}\mathbf{A} = \mathbf{L}\mathbf{U} \tag{3.8}$$

with a permutation matrix $\mathbf{P}$ and the lower $\mathbf{L}$ as well as the upper matrices $\mathbf{U}$ in triangle forms.

$$\mathbf{L} = \begin{bmatrix} 1 & \cdots & 0 \\ \vdots & \ddots & \vdots \\ l_{n1} & \cdots & 1 \end{bmatrix}, \quad \mathbf{U} = \begin{bmatrix} u_{11} & \cdots & u_{1n} \\ \vdots & \ddots & \vdots \\ 0 & \cdots & u_{nn} \end{bmatrix} \tag{3.9}$$

If $\mathbf{P} = \mathbf{I}$ the matrix $\mathbf{A}$ has a so-called LU-decomposition: $\mathbf{A} = \mathbf{L}\mathbf{U}$. The task reduces then to calculate the lower and upper matrices and invert them. Once $\mathbf{L}$ and $\mathbf{U}$ are determined, the inversion of $\mathbf{A}$ is trivial due to the triangle structure of $\mathbf{L}$ and $\mathbf{U}$.

Assuming that beside non-singularity and existing LU-decomposition, $\mathbf{A}$ is symmetrical additionally, we have $\mathbf{U} = \mathbf{D}\mathbf{L}^T$ with $\mathbf{D} = diag(d_i)$. Now we can conduct the following transformations.

$$\mathbf{A} = \mathbf{L}\mathbf{U} = \mathbf{L}\mathbf{D}\mathbf{L}^T = \underbrace{\mathbf{L}\sqrt{\mathbf{D}}}_{\tilde{\mathbf{L}}}\,\underbrace{\sqrt{\mathbf{D}}\mathbf{L}^T}_{\tilde{\mathbf{L}}^T} \tag{3.10}$$

The splitting of $\mathbf{D}$ requires that $\mathbf{A}$ is positive definite thus that $\forall d_i > 0$. The expression

$$\mathbf{A} = \tilde{\mathbf{L}}\tilde{\mathbf{L}}^T \tag{3.11}$$

is denoted as Cholesky decomposition. Therefore, the lower triangle matrices of both the Cholesky and the Gaussian method are connected via

$$\tilde{\mathbf{L}} = \mathbf{L}^T \sqrt{\mathbf{D}} \tag{3.12}$$

3.4.1.2 Iterative Methods

High resolution FEM leads to large equation systems with sparse system matrices. For this type of problems iterative equation solver are much more efficient than direct solvers. Concerning the application of iterative solver we have to distinguish between symmetrical and non-symmetrical system matrices with different solution methods. The efficiency of iterative algorithms, i.e. the reduction of iteration numbers, can be improved by the use of pre-conditioning techniques).

The last two rows of solver for symmetric problems belong to the linear steady iteration methods. The algorithms for solving non-symmetrical systems are also denoted as Krylov subspace methods.

Symmetric matrices	Non-symmetric matrices
CG	BiCG
Lanczos	CGStab
Gauss-Seidel, Jacobian, Richards	GMRES
SOR and block-iteration	CGNR

3.4.2 Non-linear Solver

In this section we present a description of selected iterative methods that are commonly applied to solve non-linear problems.

- Picard method (fixpoint iteration)
- Newton methods
- Cord slope method
- Dynamic relaxation method

All methods call for an initial guess of the solution to start but each algorithm uses a different scheme to produce a new (and hopefully closer) estimate to the exact solution. The general idea is to construct a sequence of linear sub-problems which can be solved with ordinary linear solver (see Sect. 3.4.1).

3.4.2.1 Picard Method

The general algorithm of the Picard method can be described as follows. We consider a non-linear equation written in the form

$$\mathbf{A(x)\,x - b(x)} = 0 \tag{3.13}$$

We start the iteration by assuming an initial guess $\mathbf{x}_0$ and we use this to evaluate the system matrix $\mathbf{A(x_0)}$ as well as the right-hand-side vector $\mathbf{b(x_0)}$. Thus this equation becomes linear and it can be solved for the next set of $\mathbf{x}$ values.

$$\mathbf{A(x_{k-1})\,x_k - b(x_{k-1})} = 0$$
$$\mathbf{x}_k = \mathbf{A}^{-1}(\mathbf{x}_{k-1})\,\mathbf{b}(\mathbf{x}_{k-1}) \tag{3.14}$$

Repeating this procedure we obtain a sequence of successive solutions for $\mathbf{x}_k$. During each iteration loop the system matrix and the right-hand-side vector must be updated with the previous solution. The iteration is performed until satisfactory convergence is achieved. A typical criterion is e.g.

$$\varepsilon \geq \frac{\|\,\mathbf{x}_k - \mathbf{x}_{k-1}\,\|}{\|\,\mathbf{x}_k\,\|} \tag{3.15}$$

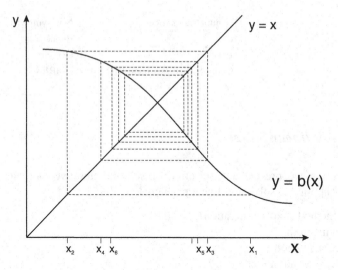

Fig. 3.4 Graphical illustration of the Picard iteration method

where ε is a user-defined tolerance criterion. For the simple case of a non-linear equation $\mathbf{x} = \mathbf{b}(\mathbf{x})$ (i.e. $\mathbf{A} = \mathbf{I}$), the iteration procedure is graphically illustrated in Fig. 3.4. To achieve convergence of the scheme it has to be guaranteed that the iteration error

$$e_k = \| \mathbf{x}_k - \mathbf{x} \| < C \| \mathbf{x}_{k-1} - \mathbf{x} \|^p = e_{k-1} \tag{3.16}$$

or, alternatively, the distance between successive solutions will reduce

$$\| \mathbf{x}_{k+1} - \mathbf{x}_k \| < \| \mathbf{x}_k - \mathbf{x}_{k-1} \|^p \tag{3.17}$$

where p denotes the convergence order of the iteration scheme. It can be shown that the iteration error of the Picard method decreases linearly with the error at the previous iteration step. Therefore, the Picard method is a first-order convergence scheme.

3.4.2.2 Newton Method

In order to improve the convergence order of non-linear iteration methods, i.e. derive higher-order schemes, the Newton-Raphson method can be employed. To describe this approach, we consider once again the non-linear equation (3.5).

$$\mathbf{R}(\mathbf{x}) = \mathbf{A}(\mathbf{x})\,\mathbf{x} - \mathbf{b}(\mathbf{x}) = 0 \tag{3.18}$$

Assuming that the residuum $\mathbf{R}(\mathbf{x})$ is a continuous function, we can develop a Taylor series expansion about any known approximate solution $\mathbf{x}_k$.

$$\mathbf{R}_{k+1} = \mathbf{R}_k + \left[\frac{\partial \mathbf{R}}{\partial \mathbf{x}}\right]_k \Delta\mathbf{x}_{k+1} + O(\Delta\mathbf{x}_{k+1}^2) \qquad (3.19)$$

Second- and higher-order terms are truncated in the following. The term $\partial\mathbf{R}/\partial\mathbf{x}$ represents tangential slopes of $\mathbf{R}$ with respect to the solution vector and it is denoted as the Jacobian matrix $\mathbf{J}$. As a first approximation we can assume $\mathbf{R}_{k+1} = 0$. Then the solution increment can be immediately calculated from the remaining terms in Eq. (3.19).

$$\Delta\mathbf{x}_{k+1} = -\mathbf{J}_k^{-1}\,\mathbf{R}_k \qquad (3.20)$$

where we have to cope with the inverse of the Jacobian. The iterative approximation of the solution vector can be computed now from the increment.

$$\mathbf{x}_{k+1} = \mathbf{x}_k + \Delta\mathbf{x}_{k+1} \qquad (3.21)$$

Once an initial guess is provided, successive solutions of $\mathbf{x}_{k+1}$ can be determined using Eqs. (3.20) and (3.21) (Fig. 3.5). The Jacobian has to re-evaluated and inverted at every iteration step, which is a very time-consuming procedure in fact. At the expense of slower convergence, the initial Jacobian $\mathbf{J}_0$ may be kept and used in the subsequent iterations. Alternatively, the Jacobian can be updated in certain iteration intervals. This procedure is denoted as modified or 'initial slope' Newton method (Fig. 3.6).

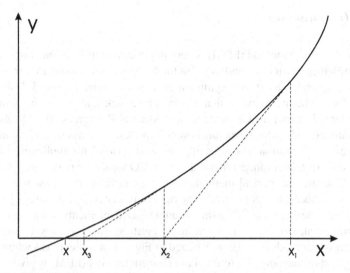

Fig. 3.5 Graphical illustration of the Newton-Raphson iteration method

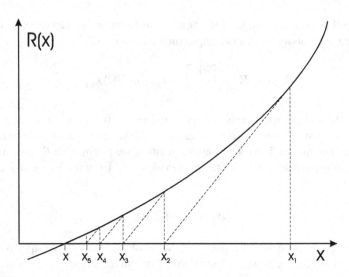

Fig. 3.6 Graphical illustration of the modified Newton-Raphson iteration method

The convergence velocity of the Newton-Raphson method is second-order. It is characterized by the expression.

$$\| \, \mathbf{x}_{k+1} - \mathbf{x} \, \| \leq C \, \| \, \mathbf{x}_k - \mathbf{x} \, \|^2 \qquad (3.22)$$

3.5 Finite Element Method

3.5.1 Introduction

The finite difference method (FDM) is very popular in numerical modeling as to its simple implementation and handling. But the FDM has limitations for representing complex geometries as it is relying on structured grids. Figure 3.7 shows an example for a fracture network that can occur in rock masses [figure source: H. Kunz, Federal Institute for Geosciences and Mineral Resources (BGR)]. Accurate representation of subsurface structure is very important to correctly understand flow, transport and deformation processes in geological systems for applications such as hazardous waste deposition, geothermal energy, CO2 and energy storage. Therefore, more sophisticated numerical methods have been developed in past to overcome limitations in accurate geometrical description if necessary. A large geometric flexibility can be achieved by using triangle-based elements such as triangles itself, tetrahedral, prismatic and pyramidal entities. An overview of numerical approximation methods have been depicted in Fig. 3.1 in a previous section. In the following, a short introduction to the finite element method (FEM) is given.

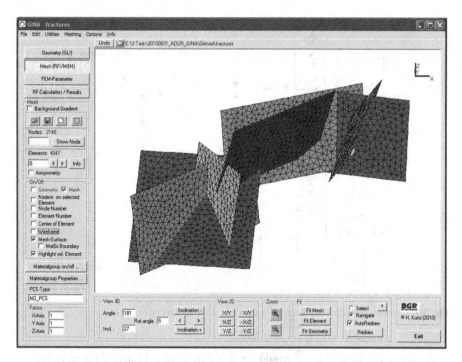

Fig. 3.7 Modeling of a fracture system in crystalline rock (Herbert Kunz, BGR)

3.5.2 Finite Element Example

To explain the finite element method we use a simple test example for steady heat conduction in a column. This exercise is based on a similar example for steady groundwater flow by Istok (1989). Based on this we can construct much more complex examples such as groundwater models for large deep aquifers systems in Saudi Arabia or in arid area like the Middle, seawater intrusion models for Oman, or investigation groundwater deterioration the subsurface of Beijing. Those examples you can find as videos on the OGS website www.opengeosys.org and in the related scientific publications. For complex systems we use the method of scientific visualization to learn about more details inside the models.

We start with a very simple example of heat conduction in a soil column (Fig. 3.8). There is no difference of the basic principle of FEM between simple and more complex examples except of computational times.

We consider a 1D steady heat conduction problem.

$$\frac{\partial}{\partial x}\left(\alpha_x \frac{\partial T}{\partial x}\right) = 0 \tag{3.23}$$

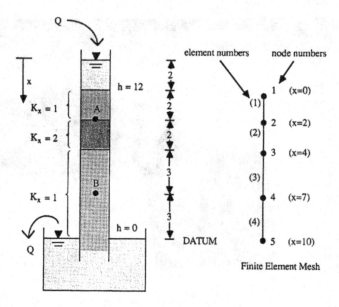

Fig. 3.8 Column model for explanation of the FEM after Istok (1989)

An approximate solution $\hat{T}$ would not fulfill the original Eq. (3.23) precisely.

$$\frac{\partial}{\partial x}\left(\alpha_x \frac{\partial \hat{T}}{\partial x}\right) = R(x) \neq 0 \qquad (3.24)$$

where $R(x)$ is the so-called residuum representing the error introduced by the numerical approximation. The residuum can be different in various grid nodes i, j having different values $R_i \neq R_j$. As an example, grid node number 3 is link to the two elements 2 and 3 (Fig. 3.8). The residuum at grid node 3, therefore, calculates from the element values as follow

$$R_3 = R_3^{(2)} + R_3^{(3)} \qquad (3.25)$$

where the exponents indicate the elements' contributions to the error. Now we can write for the residuum for each grid node i

$$R_i = \sum_{e=1}^{p} R_i^{(e)} \qquad (3.26)$$

where the index e is running over all to node i connected elements p. The element contribution to the residuum of node i is calculated as follows

$$R_i^{(e)} = \int_{x_i^e}^{x_j^e} N_i^{(e)} \left(\alpha_x^{(e)} \frac{\partial^2 \hat{T}^{(e)}}{\partial x^2}\right) dx \qquad (3.27)$$

Fig. 3.9 Interpolation function for the Galerkin method after Istok (1989)

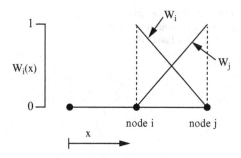

Here $N_i^{(e)} \equiv W_i(x)$ is an interpolation function on the element (e) (Fig. 3.9). The same relation we can write for the other element node j

$$R_j^{(e)} = \int_{x_i^{(e)}}^{x_j^{(e)}} N_j^{(e)} \left(\alpha_x^{(e)} \frac{\partial^2 \hat{T}^{(e)}}{\partial x^2} \right) dx \tag{3.28}$$

The linear interpolation functions for a 1D element are

$$N_i^{(e)}(x) = \frac{x_j^{(e)} - x}{L^{(e)}} \quad , \quad N_j^{(e)}(x) = \frac{x - x_i^{(e)}}{L^{(e)}} \tag{3.29}$$

The approximated field quantity T can be interpolated on a 1D finite element as follows (Fig. 3.10)

$$\hat{T}^{(e)}(x) = N_i^{(e)} T_i + N_j^{(e)} T_j$$

$$= \frac{x_j^{(e)} - x}{L^{(e)}} T_i + \frac{x - x_i^{(e)}}{L^{(e)}} T_j \tag{3.30}$$

Now we have to solve a more "serious" problem. In Eqs. (3.27) and (3.28) we have to derive second order derivatives—but the interpolation functions selected are only linear ones which cannot be derived twice. What can we do? We are using a mathematical "trick". A partial derivation of Eq. (3.27) yields.

$$\int_{x_i^e}^{x_j^e} N_i^{(e)} \left(K_x^{(e)} \frac{\partial^2 \hat{T}^{(e)}}{\partial x^2} \right) dx = - \int_{x_i^e}^{x_j^e} \alpha_x^{(e)} \frac{\partial N_i^{(e)}}{\partial x} \frac{\partial \hat{T}^{(e)}}{\partial x} dx + N_i^{(e)} \alpha_x^{(e)} \frac{\partial \hat{T}^{(e)}}{\partial x} \bigg|_{x_i^e}^{x_j^e} \tag{3.31}$$

How we can prove the above transformation of Eq. (3.31) The second term on the right hand side of Eq. (3.31)

$$N_i^{(e)} \alpha_x^{(e)} \frac{\partial \hat{T}^{(e)}}{\partial x} \bigg|_{x_i^e}^{x_j^e} \tag{3.32}$$

Fig. 3.10 Interpolated approximate solution on a 1D element after Istok (1989)

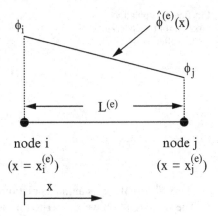

corresponds to the values on the boundary nodes x_j^e and x_i^e of the finite element (e). If it is a boundary node then the corresponding boundary condition has to be applied. Some questions before we continue:

- Which type of boundary condition is Eq. (3.32)?
- Which boundary condition do we have if the value (3.32) is equal to zero?
- What happens to the inner nodes concerning the element boundary terms?

Now we substitute the expression (3.31) into Eq. (3.27) and we yield.

$$R_i^{(e)} = \int_{x_i^e}^{x_j^e} N_i^{(e)} \left(\alpha_x^{(e)} \frac{\partial^2 \hat{T}^{(e)}}{\partial x^2} \right) dx$$

$$= -\int_{x_i^e}^{x_j^e} \alpha_x^{(e)} \frac{\partial N_i^{(e)}}{\partial x} \frac{\partial \hat{T}^{(e)}}{\partial x} dx + N_i^{(e)} \alpha_x^{(e)} \frac{\partial \hat{T}^{(e)}}{\partial x} \Big|_{x_i^e}^{x_j^e} \qquad (3.33)$$

Now we have to deal we the expression $\partial \hat{h}^{(e)} / \partial x$.

$$\frac{\partial \hat{T}^{(e)}}{\partial x} = \frac{\partial}{\partial x} \left(N_i^{(e)} T_i + N_j^{(e)} T_j \right) = \frac{\partial N_i^{(e)}}{\partial x} T_i + \frac{\partial N_j^{(e)}}{\partial x} T_j \qquad (3.34)$$

After inserting of the interpolation functions we receive.

$$\frac{\partial \hat{h}^{(e)}}{\partial x} = \frac{1}{L^{(e)}} (-h_i + h_j) \qquad (3.35)$$

For the derivations of the linear interpolation functions we have.

$$\frac{\partial N_i^{(e)}}{\partial x} = \frac{\partial}{\partial x}\left(\frac{x_j^{(e)} - x}{L^{(e)}}\right) = -\frac{1}{L^{(e)}} \tag{3.36}$$

$$\frac{\partial N_j^{(e)}}{\partial x} = \frac{\partial}{\partial x}\left(\frac{x - x_i^{(e)}}{L^{(e)}}\right) = \frac{1}{L^{(e)}} \tag{3.37}$$

If we insert the above relation into Eq. (3.33) we yield for the residuum.

$$R_i^{(e)} = \int_{x_i^e}^{x_j^e} \alpha_x^{(e)} \left(-\frac{1}{L^{(e)}}\right)\left(\frac{1}{L^{(e)}}\right)(-T_i + T_j)$$

$$= \frac{\alpha_x^{(e)}}{L^{(e)}}(T_i - T_j) \tag{3.38}$$

In the same way we become.

$$R_j^{(e)} \frac{\alpha_x^{(e)}}{L^{(e)}}(-T_i + T_j) \tag{3.39}$$

Both Eqs. (3.38) and (3.39) can be written in a matrix form.

$$\left\{\begin{array}{c} R_i^{(e)} \\ R_j^{(e)} \end{array}\right\} = \frac{\alpha_x^{(e)}}{L^{(e)}} \underbrace{\left[\begin{array}{cc} +1 & -1 \\ -1 & +1 \end{array}\right]}_{2\times 2} \left\{\begin{array}{c} T_i \\ T_j \end{array}\right\} \tag{3.40}$$

with the conductivity matrix:

$$[K^{(e)}] = \frac{K_x^{(e)}}{L^{(e)}} \underbrace{\left[\begin{array}{cc} +1 & -1 \\ -1 & +1 \end{array}\right]}_{2\times 2} \tag{3.41}$$

Based on the element geometries $L^{(e)}$ (Fig. 3.8) we obtain the following element matrices.

$$[K^{(1)}] = \left[\begin{array}{cc} +1/2 & -1/2 \\ -1/2 & +1/2 \end{array}\right] \quad , \quad [K^{(2)}] = \left[\begin{array}{cc} +1 & -1 \\ -1 & +1 \end{array}\right]$$

$$[K^{(1)}] = \left[\begin{array}{cc} +1/3 & -1/3 \\ -1/3 & +1/3 \end{array}\right] \quad , \quad [K^{(2)}] = \left[\begin{array}{cc} +1/3 & -1/3 \\ -1/3 & +1/3 \end{array}\right]$$

$$\tag{3.42}$$

Finally we assemble the equation system based on combining the element matrices.

$$\{R\} = [K]\{T\} = 0 \tag{3.43}$$

$$\{R\} = \begin{bmatrix} R_1 \\ R_2 \\ R_3 \\ R_4 \\ R_5 \end{bmatrix} \quad , \quad \{T\} = \begin{bmatrix} T_1 \\ T_2 \\ T_3 \\ T_4 \\ T_5 \end{bmatrix} \tag{3.44}$$

$$[K] = [K^{(1)}] + [K^{(2)}] + [K^{(3)}] + [K^{(4)}] \tag{3.45}$$

$$[K] = \begin{bmatrix} 1/2 & -1/2 & 0 & 0 & 0 \\ -1/2 & 1+1/2 & -1 & 0 & 0 \\ 0 & -1 & 1+1/3 & -1/3 & 0 \\ 0 & 0 & -1/3 & 1/3+1/3 & -1/3 \\ 0 & 0 & 0 & -1/3 & 1/3 \end{bmatrix}$$

$$= \begin{bmatrix} 1/2 & -1/2 & 0 & 0 & 0 \\ -1/2 & 3/2 & -1 & 0 & 0 \\ 0 & -1 & 4/3 & -1/3 & 0 \\ 0 & 0 & -1/3 & 2/3 & -1/3 \\ 0 & 0 & 0 & -1/3 & 1/3 \end{bmatrix}$$

$$\begin{bmatrix} 1/2 & -1/2 & 0 & 0 & 0 \\ -1/2 & 3/2 & -1 & 0 & 0 \\ 0 & -1 & 4/3 & -1/3 & 0 \\ 0 & 0 & -1/3 & 2/3 & -1/3 \\ 0 & 0 & 0 & -1/3 & 1/3 \end{bmatrix} \begin{Bmatrix} T_1 \\ T_2 \\ T_3 \\ T_4 \\ T_5 \end{Bmatrix} = \begin{Bmatrix} 0 \\ 0 \\ 0 \\ 0 \\ 0 \end{Bmatrix} \tag{3.46}$$

3.5.3 Time Derivatives for Transient Processes

In order to build the time derivative for transient processes we simply rely on finite differences, i.e.

$$\frac{\partial T}{\partial t} \approx \frac{T^{n+1} - T^n}{t^{n+1} - t^n} \tag{3.47}$$

3.6 Exercises

3.6.1 Solution Procedure

1. Give examples of approximation methods for solving differential equations.
2. What are the two basic steps of the solution procedure for discrete approximation methods ?

3.6.2 Theory of Discrete Approximation

3. Explain the term convergence of an approximate solution for a PDE. Give a mathematical definition for that.
4. Explain the term consistency of an approximation scheme for a PDE. Give a mathematical definition for that.
5. Explain the term stability of an approximate solution for a PDE. What general methods for stability analysis do you know ?
6. Explain the relationships between the terms convergence, consistency, and stability using Fig. (3.3).
7. What does the Lax equivalence theorem postulate ?
8. What are the three analysis steps for discrete approximation schemes ?

3.6.3 Solution Process

9. Using the Newton-Raphson method solve the following set of non-linear equations:

$$f_1(x_1, x_2) = x_1^2 + x_2^2 - 5 = 0$$
$$f_2(x_1, x_2) = x_1 + x_2 - 1 = 0$$

3.6.4 Finite Element Method

10. Explain the advantages and disadvantages of the finite element method (FEM).
11. Which geometric element types can be represented by the FEM ?
12. What is a residuum?
13. What is the native boundary condition of FEM ?
14. Calculate the element conductivity matrices for the following finite element mesh.

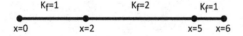

Chapter 4
Heat Transport Exercises

After the theory lectures touching aspects of continuum mechanics of porous media, thermodynamics and an introduction of numerical methods, now we conduct five exercises on heat transport simulation using OGS—so called TH (thermo-hydraulic) processes.

4.1 Exercises Schedule

More benchmarks and examples can be obtained from the existing benchmark book volume 1 and volume 2 (Kolditz et al. 2012, 2015a) as well as the OGS community webpage www.opengeosys.org.

4.2 OGS File Concept

Table 4.1 gives also the directory structure for the exercises on heat transport problems. The numerical simulation with *OGS* relies on file-based model setups, which means each model needs different input files that contain information on specific aspects of the model. All the input files share the same base name but have a unique file ending, with which the general information of the file can already be seen. For example, a file with ending .pcs provides the information of the process involved in the simulation such as groundwater flow or Richards flow; whereas in a file with ending .ic the initial condition of the model can be defined. Table 4.2 gives an overview and short explanations of the *OGS* input files needed for one of the benchmarks.

© Springer International Publishing Switzerland 2016
N. Böttcher et al., *Geoenergy Modeling I*, SpringerBriefs in Energy,
DOI 10.1007/978-3-319-31335-1_4

Table 4.1 Exercises overview

Unit	Author	Contents
OGS-CES-I-E06	Böttcher	Heat conduction in a semi-finite domain
OGS-CES-I-E07	Watanabe	Heat flux through a layered medium
OGS-CES-I-E08	Böttcher	Heat transport in a porous medium
OGS-CES-I-E09	Böttcher	Heat transport in a porous-fractured medium
OGS-CES-I-E10	Watanabe	Heat convection in a porous medium

Table 4.2 *OGS* input files for heat transport problems

Object	File	Explanation
GEO	`file.gli`	System geometry
MSH	`file.msh`	Finite element mesh
PCS	`file.pcs`	Process definition
NUM	`file.num`	Numerical properties
TIM	`file.tim`	Time discretization
IC	`file.ic`	Initial conditions
BC	`file.bc`	Boundary conditions
ST	`file.st`	Source/sink terms
MFP	`file.mfp`	Fluid properties
MSP	`file.mfp`	Solid properties
MMP	`file.mmp`	Medium properties
OUT	`file.out`	Output configuration

The basic structure and concept of an input file is illustrated in the examples below using listings (e.g. Listing 4.1). As we can see, an input file begins with a main keyword which contains sub keywords with corresponding parameter values. An input file ends with the keyword #STOP, everything written after file input terminator #STOP is unaccounted for input. Please also refer to the *OGS* input file description in Appendix and the keyword description to the *OGS* webpage (http://www.opengeosys.org/help/documentation).

4.3 Heat Conduction in a Semi-Infinite Domain

4.3.1 Definition

We consider a diffusion problem in a 1D half-domain which is infinite in one coordinate direction ($z \to \infty$) (Fig. 4.1). The benchmark test was developed and provided by Norbert Böttcher.

Fig. 4.1 Model domain

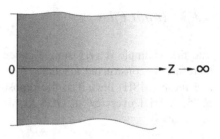

Table 4.3 Solid phase material properties

Symbol	Parameter	Value	Unit
ρ	Density	2500	$\mathrm{kg\,m^{-3}}$
c	Heat capacity	1000	$\mathrm{J\,kg^{-1}\,K^{-1}}$
λ	Thermal conductivity	3.2	$\mathrm{W\,m^{-1}\,K^{-1}}$

Fig. 4.2 Spatial discretisation of the numerical model

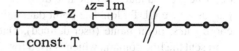

4.3.2 Solution

4.3.2.1 Analytical Solution

The analytical solution for the 1D linear heat conduction equation (2.19) is

$$T(t, z) = T_0 \mathrm{erfc}\left(\frac{z}{\sqrt{4\alpha t}}\right), \qquad (4.1)$$

where T_0 is the initial temperature. The boundary conditions are $T(z = 0) = 1$ and $T(z \to \infty) = 0$.

The material properties for this model setup are given in Table 4.3.

4.3.2.2 Numerical Solution

The numerical model consists of 60 line elements connected by 61 nodes along the z-axis (Fig. 4.2). The distances of the nodes Δz is 1 m. At $z = 0$ m there is a constant temperature boundary condition.

The *Neumann* stability criteria has to be restrained so that the temperature gradient can not be inverted by diffusive fluxes. Using (4.2) the best time step can be estimated by

$$\mathrm{Ne} = \frac{\alpha \Delta t}{(\Delta z)^2} \leq \frac{1}{2}. \qquad (4.2)$$

With $\Delta z = 1$ m and $\alpha = 1.28 \cdot 10^{-6}\,\mathrm{m^2\,s^{-1}}$ the outcome for the time step is $\Delta t \leq 390625$ s or 4.5 days, respectively.

4.3.3   Input Files

The first example is very simple concerning geometry, mesh and processes, and therefore, is constructed manually. We recommend starting with geometry (GLI) and mesh (MSH) files. The file repository is www.opengeosys.org/tutorials/cesi/e06. A brief overview of OGS keywords used in this tutorial can be found in Appendix B.

4.3.3.1   GLI: Geometry

Geometry (GLI) files contain data about geometric entities such as points, polylines, surfaces, and volumes. We define six points (#POINTS), where point 0 and 1 describe the boundaries of the line domain. Points 2–6 are specified for data output purposes (see OUT file). The point data are: point number (starting with 0), x, y, z coordinates, point name (user-defined). One polyline is given (#POLYLINE) to represent the line domain, which is used for data output again (see OUT file). Do not forget closing the file with #STOP to finish data input. After the stop keyword you can write your comments etc. You can load the GLI file using the OGS DataExplorer for data visualization.

Listing 4.1   GLI input file

```
#POINTS                    // points keyword
0  0  0  0  $NAME POINT0   // point number | x | y | z | point name
1  0  0  60
2  0  0  1  $NAME POINT2
3  0  0  2  $NAME POINT3
4  0  0  10 $NAME POINT4
5  0  0  20 $NAME POINT5
#POLYLINE                  // polyline keyword
 $NAME                     // polyline name subkeyword
  ROCK                     // polyline name
 $POINTS                   // polyline points subkeyword
  0                        // point of polyline
  1                        // dito
#STOP                      // end of input data
```

4.3.3.2   MSH: Finite Element Mesh

Mesh (MSH) files contain data about the finite element mesh(es) such as nodes and elements. Node data ($NODES) are the grid node number as well as the coordinates x,y,z (you always have to give all three coordinates even for 1D or 2D examples as OGS is "thinking" in 3D). Element data ($ELEMENTS) contain more information: the element number (beginning from 0, must be unique), the associated material group (see MMP file), the geometric element type (here a linear line element with two element nodes), and finally the element node number, i.e. the grid nodes forming the element. A finite element is a topological item, therefore orientation and node numbering matters.

Listing 4.2 MSH input file

```
#FEM_MSH  // file/object keyword
$NODES    // node subkeyword
61        // number of grid nodes
0 0 0 0   // node number x y z
1 0 0 1   // dito
...
59     0       0       59
60     0       0       60
$ELEMENTS     // element subkeyword
60            // number of elements
0      0      line    0       1  // element number | material group number |
    element type | element node numbers
1      0      line    1       2
...
58     0      line    58      59
59     0      line    59      60
#STOP         // end of data part
```

4.3.3.3 PCS: Process Definition

Process (PCS) files (Table 4.2) specify the physico-biochemical process being simulated. OGS is a fully coupled THMC (thermo-hydro-mechanical-chemical) simulator, therefore, a large variety of process combination is available with subsequent dependencies for OGS objects.

Listing 4.3 PCS input file

```
#PROCESS          // file/object keyword
 $PCS_TYPE        // process subkeyword
  HEAT_TRANSPORT  // specified process
#STOP             // end of input data
```

4.3.3.4 NUM: Numerical Properties

The next set of two files (NUM and TIM) are specifying numerical parameters, e.g. for spatial and temporal discretization as well as parameters for equation solvers.

Numerics (NUM) files (Table 4.2) contain information for equation solver, time collocation, and Gauss points. The process subkeyword ($PCS_TYPE) specifies the process to whom the numerical parameters belong to (i.e. HEAT_TRANSPORT). The linear solver subkeyword ($LINEAR_SOLVER) then determines the parameters for the linear equation solver (Table 4.4).

Listing 4.4 NUM input file

```
#NUMERICS         // file/object keyword
 $PCS_TYPE        // process subkeyword
  HEAT_TRANSPORT  // specified process
 $LINEAR_SOLVER   // linear solver subkeyword
; method error_tolerance max_iterations theta precond storage
   2      0 1.e-012        1000          1.0   100     4 // solver parameters
```

Table 4.4 $LINEAR_SOLVER subkeyword data

Parameter	Number	Meaning
Method	2	Iterative CG solver: SpBiCG
Error method	0	Absolute error
Error tolerance	10^{-12}	Error tolerance due error method
Maximum iterations	1000	Maximum number of solver iterations
Theta	1.0	Collocation parameter
Preconditioner	100	Preconditioner method
Storage	4	Storage model

```
$ELE_GAUSS_POINTS // Gauss points subkeyword
  2               // number of Gauss points
#STOP             // end of input data
```

4.3.3.5   TIM: Time Discretization

Time discretization (TIM) files (Table 4.2) specify the time stepping schemes for
related processes. The process subkeyword ($PCS_TYPE) specifies the process to
whom the time discretization parameters belong to (i.e. HEAT_TRANSPORT).

Listing 4.5 TIM input file

```
#TIME_STEPPING      // file/object keyword
  $PCS_TYPE         // process subkeyword
    HEAT_TRANSPORT  // specified process
  $TIME_STEPS       // time steps subkeyword
    1000 390625     // number of times steps | times step length
  $TIME_END         // end time subkeyword
    1E99            // end time value
  $TIME_START       // starting time subkeyword
    0.0             // starting time value
#STOP               // end of input data
```

4.3.3.6   IC: Initial Conditions

The next set of files (IC/BC/ST) are specifying initial and boundary conditions as
well as source and sink terms for related processes. Initial conditions (IC), boundary
conditions (BC) and source/sink terms (ST) are node related properties.

Listing 4.6 IC input file

```
#INITIAL_CONDITION
  $PCS_TYPE
    HEAT_TRANSPORT
  $PRIMARY_VARIABLE
    TEMPERATURE1
  $GEO_TYPE
    DOMAIN
```

```
$DIS_TYPE
  CONSTANT    0
#INITIAL_CONDITION
 $PCS_TYPE
  HEAT_TRANSPORT
 $PRIMARY_VARIABLE
  TEMPERATURE1
 $GEO_TYPE
  POINT POINT0
 $DIS_TYPE
  CONSTANT    1
#STOP
```

4.3.3.7   BC: Boundary Conditions

The boundary conditions (BC) file (Table 4.2) assigns the boundary conditions to the model domain. The following example applies a constant Dirichlet boundary condition value 1.0 for the heat transport process for the primary variable temperature at the point with name POINT0. Note that BC objects are linked to geometry objects (here POINT).

Listing 4.7 BC input file

```
#BOUNDARY_CONDITION // boundary condition keyword
 $PCS_TYPE            // process type subkeyword
  HEAT_TRANSPORT      // specified process
 $PRIMARY_VARIABLE    // primary variable subkeyword
  TEMPERATURE1        // specified primary variable
 $GEO_TYPE            // geometry type subkeyword
  POINT POINT0        // specified geometry type | geometry name
 $DIS_TYPE            // boundary condition type subkeyword
  CONSTANT    1       // boundary condition type | value
#STOP
```

4.3.3.8   ST: Source/Sink Terms

• not required here

4.3.3.9   MFP: Fluid Properties

The fluid properties (MFP) file (Table 4.2) defines the material properties of the fluid phase(s). For multi-phase flow models we have multiple fluid properties objects. It contains physical parameters such as fluid density ρ^f, dynamic fluid viscosity μ such as heat capacity c^f, and thermal conductivity λ^f. The first parameter for the material properties is the material model number.

Listing 4.8 MFP input file

```
#FLUID_PROPERTIES
 $DENSITY
  1 1000.0
 $VISCOSITY
  1 0.0
 $SPECIFIC_HEAT_CAPACITY
  1 0.0
 $HEAT_CONDUCTIVITY
  1 0.0
#STOP
```

4.3.3.10   MSP: Solid Properties

The solid properties (MSP) file (Table 4.2) defines the material properties of the solid phase. It contains physical parameters such as solid density ρ^s, thermophysical parameters such as thermal expansion coefficient β_T^s, heat capacity c^s, and thermal conductivity λ^s. The first parameter for the material properties is the material model number.

Listing 4.9 MSP input file

```
#SOLID_PROPERTIES
 $DENSITY
  1 2500
 $THERMAL
  EXPANSION:
   1 0
  CAPACITY:
   1 1000
  CONDUCTIVITY:
   1 3.2
#STOP
```

4.3.3.11   MMP: Porous Medium Properties

The medium properties (MMP) file (Table 4.2) defines the material properties of the porous medium for all processes (single continuum approach). It contains geometric properties related to the finite element dimension (geometry dimension and area) as well as physical parameters such as porosity n, storativity S_0, tortuosity τ, permeability tensor $\mathbf{k}$, and heat dispersion parameters α_L, α_T. The first parameter for the material properties is the material model number.

Listing 4.10 MMP input file

```
#MEDIUM_PROPERTIES
 $GEOMETRY_DIMENSION
  1
 $GEOMETRY_AREA
  1.0
 $POROSITY
```

Table 4.5 #OUTPUT subkeywords and parameters

Subkeyword	Parameter	Meaning
$PCS_TYPE	Process name	HEAT_TRANSPORT
$NOD_VALUES	Nodal values to be output	TEMPERATURE1
$GEO_TYPE	Geometric type and name	POLYLINE ROCK
$TIM_TYPE	Temporal type and parameters	STEPS 1

```
 1   0.10
$STORAGE
 1   0.0
$TORTUOSITY
 1   1.000000e+000
$PERMEABILITY_TENSOR
 ISOTROPIC    1.0e-15
$HEAT_DISPERSION
 1   0.000000e+000 0.000000e+000
#STOP
```

4.3.3.12 OUT: Output Parameters

The output (OUT) file (Table 4.2) specifies the output objects for related processes. Output objects are related to a specific process and define the output values on which geometry and at which times. Table 4.5 lists the output subkeywords and related parameters.

The following output file contains two output objects, first, output of data along a polyline at each time step and, second, output of data in a point at each time step.

Listing 4.11 OUT input file

```
#OUTPUT             // output keyword
 $PCS_TYPE          // process subkeyword
  HEAT_TRANSPORT    // specified process
 $NOD_VALUES        // nodal values subkeyword
  TEMPERATURE1      // specified nodal values
 $GEO_TYPE          // geometry type subkeyword
  POLYLINE ROCK     // geometry type and name
 $TIM_TYPE          // output times subkeyword
  STEPS 1           // output methods and parameter
#OUTPUT
 $PCS_TYPE
  HEAT_TRANSPORT
 $NOD_VALUES
  TEMPERATURE1
 $GEO_TYPE
  POINT POINT2
 $TIM_TYPE
  STEPS 1
#STOP
```

4.3.4   Run the Simulation

After having all input files completed you can run your first simulation. Download
the latest version from the OGS website http://www.opengeosys.org/resources/
downloads. It is recommended to copy the executable `ogs.exe` into the working
directory, where the input files are located. Start the Windows CMD console
application:

1. `c:\\windows\\systems32\\cmd.exe`
2. Figs. 4.3, 4.4, 4.5, 4.6, 4.7

Fig. 4.3 Starting Windows
CMD console application

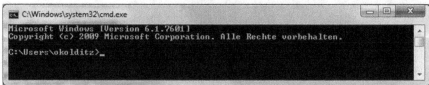

Fig. 4.4 After starting Windows CMD the console application will appear

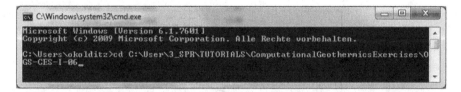

Fig. 4.5 Change to your working directory

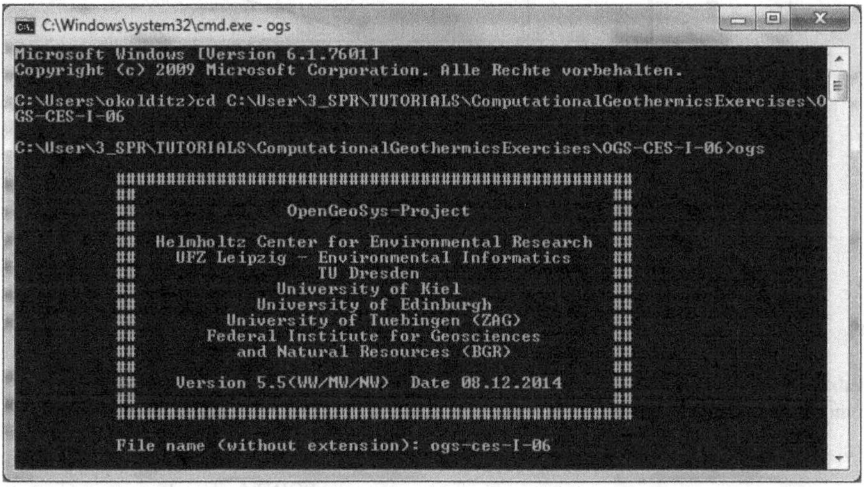

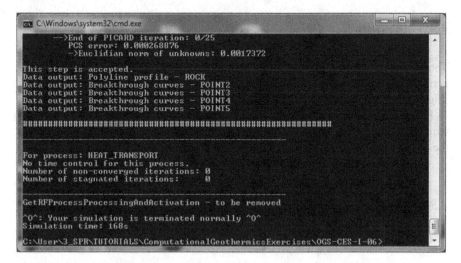

Fig. 4.6 Execute ogs from the console application and type the name of the input file (no file extension)

Fig. 4.7 The simulation takes about 160 s and output files are available

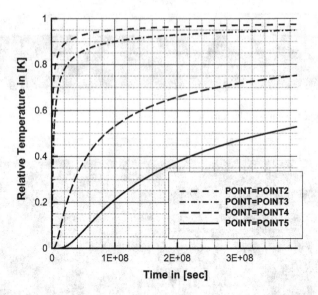

Fig. 4.8 Temperature evolution in the specified observation points (see Table 4.6)

Table 4.6 Output points for
temporal breakthrough curves

POINT	Coordinates (z in [m])
POINT2	$z = 1$
POINT3	$z = 2$
POINT4	$z = 10$
POINT5	$z = 20$

4.3.5   Results

The results are shown in two different representations. First we show the temperature evolution in the specified observation points (see OUT file): POINT2-5 (Fig. 4.8).

Figure 4.9 shows the comparison of the analytical solution of (4.1) and the numerical simulation results. The temperature distribution is demonstrated along the model domain after 2 months, 1 year, 2 years and 4 years.

4.3.6   Further Exercises

1. Set an output point at $z =5$m.

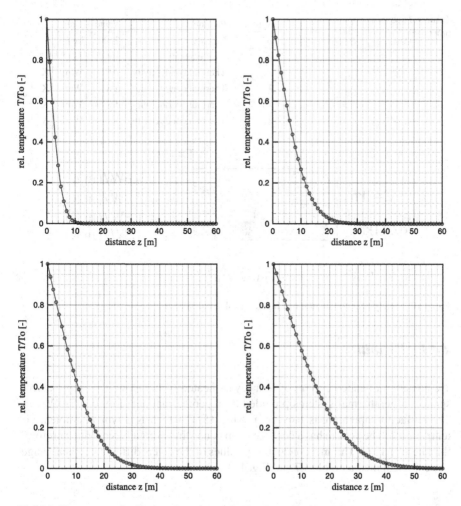

Fig. 4.9 Temperature distribution along the z-axis after 2 months, 1 year, 2 years and 4 years (*from top left to bottom right*)

4.4 Heat Flow Through a Layered Porous Medium

The second exercise is about steady heat conduction in a heterogeneous (layered) geological system. Thermal conductivities are different in each layer. A heat flux boundary condition is applied from the bottom boundary to represent a terrestrial heat flux.

In comparison to the previous exercise, the following new OGS modeling features are introduced and explained in detail:

- Dealing with heterogeneous media,
- Applying flux boundary conditions (Neumann type boundary conditions).

Table 4.7 Material
properties

Symbol	Parameter	Value	Unit
ρ	Density	2500	$\mathrm{kg\,m^{-3}}$
c	Heat capacity	1000	$\mathrm{J\,kg^{-1}\,K^{-1}}$
λ_1	Thermal conductivity	3.6	$\mathrm{W\,m^{-1}\,K^{-1}}$
λ_2	Thermal conductivity	2.8	$\mathrm{W\,m^{-1}\,K^{-1}}$
λ_3	Thermal conductivity	1.8	$\mathrm{W\,m^{-1}\,K^{-1}}$
λ_4	Thermal conductivity	4.3	$\mathrm{W\,m^{-1}\,K^{-1}}$

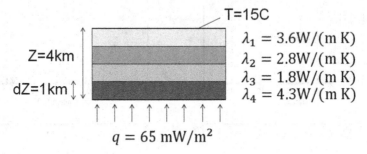

Fig. 4.10 Steady heat conduction in a layered geological system

4.4.1   Definition

We consider a bedded system of four layers heated from below. This setup corresponds to geothermal systems provided by a geothermal heat flux of $65\,\mathrm{mW/m^2}$. The surface temperature is fixed at a value $15\,^\circ\mathrm{C}$ which is a typical ambient air temperature. The system has a depth of 4 km consisting of four layers with a uniform thickness of 1 km. Thermal conductivity values of the layers differ in a typical range of geological media (Table 4.7 and Fig. 4.10).

4.4.2   Input Files

The second example is very simple concerning geometry, mesh and processes, as well and therefore, we construct it manually. After a very detailed description of the input files of the first example, we will highlight in the following only the new features used in the OGS input files. The file repository is www.opengeosys.org/tutorials/ces-i/e07. A brief overview of OGS keywords used in this tutorial can be found in Appendix B. Visit the OGS web-documentation http://www.opengeosys.org/help/documentation/ for more details.

4.4.2.1 GLI: Geometry

The geometry points (0–4) indicate the position of the layer interfaces as well as the zero elevation surface position. The layer thickness is 1000 m each, a polyline PLY_0 from point 0 to 4 is defined for the vertical profile.

Listing 4.12 GLI input file

```
#POINTS
0 0 0 0 $NAME POINT0
1 0 0 -1000 $NAME POINT1
2 0 0 -2000 $NAME POINT2
3 0 0 -3000 $NAME POINT3
4 0 0 -4000 $NAME POINT4
#POLYLINE
 $NAME
  PLY_0
 $POINTS
  0
  4
#STOP
```

See the OGS web-documentation http://www.opengeosys.org/help/documen tation/geometry-file for more input details.

4.4.2.2 MSH: Finite Element Mesh

The mesh contains just 5 nodes and 4 line elements.

Listing 4.13 MSH input file

```
#FEM_MSH
 $NODES
  5
0 0 0 0
1 0 0 -1000
2 0 0 -2000
3 0 0 -3000
4 0 0 -4000
 $ELEMENTS
  4
0 0 line 0 1
1 1 line 1 2
2 2 line 2 3
3 3 line 3 4
#STOP
```

4.4.2.3 PCS: Process Definition

The selected process type is heat transport and the temporal scheme is steady-state, i.e. no transient processes are considered. Using the steady-state option makes the numerical simulation for steady-state processes very efficient, because you do not have to evolve through transient simulations towards the steady-state solution.

The OGS steady-state scheme solves the steady-state equation, i.e. transient terms are neglected—but be aware this trick will only work for linear and solely processes.

Listing 4.14 PCS input file

```
#PROCESS
 $PCS_TYPE
   HEAT_TRANSPORT
 $TIM_TYPE ; temporal scheme
   STEADY  ; steady state simulation
#STOP
```

4.4.2.4   NUM: Numerical Properties

For simulating only heat transport with constant material properties, we need only a configuration for a linear solver.

Listing 4.15 NUM input file

```
#NUMERICS
 $PCS_TYPE
  HEAT_TRANSPORT
 $LINEAR_SOLVER
; method error_tolerance max_iterations theta precond storage
  2      6 1.e-010       3000           1.0   1       4
#STOP
```

4.4.2.5   TIM: Time Discretization

For the steady-state solution we only need a single time step. The time step length for steady-state simulations is arbitrary.

Listing 4.16 TIM input file

```
#TIME_STEPPING
 $PCS_TYPE
   HEAT_TRANSPORT
 $TIME_STEPS
   1 1
 $TIME_END
   1
 $TIME_START
   0
#STOP
```

4.4.2.6   IC: Initial Conditions

We set a constant zero ($T = 0\,°C$) initial condition everywhere within the modeling domain.

Listing 4.17 IC input file

```
#INITIAL_CONDITION
 $PCS_TYPE
  HEAT_TRANSPORT
 $PRIMARY_VARIABLE
  TEMPERATURE1
 $GEO_TYPE
  DOMAIN
 $DIS_TYPE
  CONSTANT 0
#STOP
```

4.4.2.7 BC: Boundary Conditions

The upper boundary (surface point at POINT0) is a mean surface temperature value $T = 15\,°C$.

Listing 4.18 BC input file

```
#BOUNDARY_CONDITION
 $PCS_TYPE
  HEAT_TRANSPORT
 $PRIMARY_VARIABLE
  TEMPERATURE1
 $GEO_TYPE
  POINT POINT0
 $DIS_TYPE
  CONSTANT   15
#STOP
```

4.4.2.8 ST: Source/Sink Terms

A heat flux of $q_T = 0.065\,W/m^2$ is prescribed at the bottom (bottom point at POINT4)

Listing 4.19 ST input file

```
#SOURCE_TERM
 $PCS_TYPE
  HEAT_TRANSPORT
 $PRIMARY_VARIABLE
  TEMPERATURE1
 $GEO_TYPE
  POINT POINT4
 $DIS_TYPE
  CONSTANT_NEUMANN 0.065
#STOP
```

4.4.2.9   MFP: Fluid Properties

Typical fluid properties of water are configured, although it does not affect the
simulation because we assume the matrix porosity is zero, i.e. no fluid present in
the system.

Listing 4.20 MFP input file

```
#FLUID_PROPERTIES
 $DENSITY
  1 1000
 $VISCOSITY
  1 0.001
 $SPECIFIC_HEAT_CAPACITY
  1 4680
 $HEAT_CONDUCTIVITY
  1 0.6
#STOP
```

4.4.2.10   MSP: Solid Properties

The solid properties (thermal conductivity of the solid phase) differ for each layer.
Compare the layer numbers with the material group number in the MSH file.

Listing 4.21 MSP input file

```
; layer 1
#SOLID_PROPERTIES
 $DENSITY
   1 2600
 $THERMAL
   CAPACITY
   1 850
   CONDUCTIVITY
   1 3.6
; layer 2
#SOLID_PROPERTIES
 $DENSITY
   1 2600
 $THERMAL
   CAPACITY
   1 850
   CONDUCTIVITY
   1 2.8
; layer 3
#SOLID_PROPERTIES
 $DENSITY
   1 2600
 $THERMAL
   CAPACITY
   1 850
   CONDUCTIVITY
   1 1.8
; layer 4
#SOLID_PROPERTIES
 $DENSITY
   1 2600
```

```
$THERMAL
  CAPACITY
  1 850
  CONDUCTIVITY
  1 4.3
#STOP
```

4.4.2.11 MMP: Porous Medium Properties

The heterogeneous model consists of 4 layers with varying material properties. For each layer we need to define a corresponding material group. Therefore, the MMP file contains 4 MEDIUM_PROPERTIES objects even though the properties are only different for the solid phase of the porous medium layer. To simplify the exercise, we assume there is no fluid present in the system, i.e. porosity is zero. Properties of the porous medium are determined solely by its solid phase.

Listing 4.22 MMP input file

```
; layer 1
#MEDIUM_PROPERTIES
 $GEOMETRY_DIMENSION
  1
 $GEOMETRY_AREA
  1.0
 $POROSITY
  1 0.0
; layer 2
#MEDIUM_PROPERTIES
 $GEOMETRY_DIMENSION
  1
 $GEOMETRY_AREA
  1.0
 $POROSITY
  1 0.0
; layer 3
#MEDIUM_PROPERTIES
 $GEOMETRY_DIMENSION
  1
 $GEOMETRY_AREA
  1.0
 $POROSITY
  1 0.0
; layer 4
#MEDIUM_PROPERTIES
 $GEOMETRY_DIMENSION
  1
 $GEOMETRY_AREA
  1.0
 $POROSITY
  1 0.0
#STOP
```

4.4.2.12   OUT: Output Parameters

The output of simulation results is along the vertical polyline PLY_0.

Listing 4.23 OUT input file

```
#OUTPUT
 $NOD_VALUES
   TEMPERATURE1
 $GEO_TYPE
   POLYLINE PLY_0
 $DAT_TYPE
   TECPLOT
 $TIM_TYPE
   STEPS 1
#STOP
```

4.4.2.13   File Repository

The file repository is www.opengeosys.org/tutorials/ces-i/e07

4.4.3   Run the Simulation

After having all input files completed you can run the simulation. Download the latest version from the OGS website http://www.opengeosys.org/resources/downloads. It is recommended to copy the executable ogs.exe into the working directory, where the input files are located.

4.4.3.1   OGS-Tips

- Wrong file formats: OGS is a multi-platform code running for Windows, Linux, Mac operation systems (OS). If you take input files from different platforms you may to convert input files. If you receive an error message from OGS (Fig. 4.11) you need to convert the input files using e.g. the unix2dos or vice-versa utilities.
- Using batch files: For longer simulation runs or producing log files you may use batch files for executing simulations. The following listing shows the instruction of the e2.bat script for running the e2 example and writing a log file to e2log.txt.

Listing 4.24 Batch script

```
..\bin\ogs e2 > e2log.txt
```

Fig. 4.11 OGS error message for wrong file formats from different platforms

4.4.4 Results

The results for the vertical temperature distribution are shown in Fig. 4.12. The surface temperature is fixed to a value of $T = 15\,°C$. What will happen with the bottom temperature if the geothermal heat flux will change?

4.4.5 Further Exercises

1. Mesh refinement
2. The time step length for steady-state simulations is arbitrary (exercise).
3. What will happen with the bottom temperature if the geothermal heat flux will change?

4.5 Heat Transport in a Porous Medium

This example is based on the benchmark exercise introduced by Norbert Böttcher to the OGS benchmark collection (Böttcher et al. 2012).

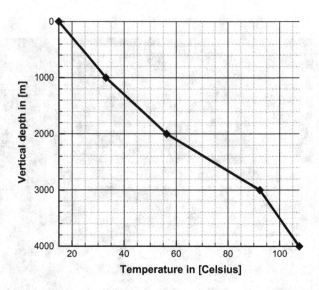

Fig. 4.12 Vertical temperature distribution

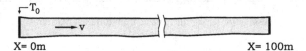

Fig. 4.13 A fully saturated fracture with flowing water and a constant temperature at the left border

Table 4.8 Model parameters

Symbol	Parameter	Value	Unit
ρ^l	Density of water	1000	kg m^{-3}
c^l	Heat capacity of water	4000	$\text{J kg}^{-1}\,\text{K}^{-1}$
λ^l	Thermal conductivity of water	0.6	$\text{W m}^{-1}\,\text{K}^{-1}$
v_x	Water velocity	$3 \cdot 10^{-7}$	m s^{-1}
L	Fracture length	100	m

4.5.1 Definition

This problem shows 1D heat transport by advection and diffusion in a $100\,m$ long fracture. The fracture is fully saturated with water, flowing with constant velocity. There is no rock matrix around the fracture considered which could store heat (this will be examined in the next example). Figure 4.13 depicts the model set-up.

The fracture is described as a porous medium with a porosity of $n = 1.0$, so that no solid material influences the heat transport process. The properties of the fluid are in Table 4.8.

These values cause a diffusivity constant for water of $\alpha = 1.5 \cdot 10^{-7} \, \text{m}^2 \, \text{s}^{-1}$. The groundwater velocity in the fracture is $v_x = 3.0 \cdot 10^{-7} \, \text{m} \, \text{s}^{-1}$.

4.5.2 Solution

For 1D-advective/diffusive transport, an analytical solution is given by Ogata and Banks (1961) as

$$T(x,t) = \frac{T_0}{2}\left(\text{erfc}\frac{x - v_x \cdot t}{\sqrt{4\alpha t}} + \exp\left(\frac{v_x \cdot x}{\alpha}\right)\text{erfc}\frac{x + v_x \cdot t}{\sqrt{4\alpha t}}\right), \tag{4.3}$$

where T_0 is the constant temperature at $x = 0$, v_x is the groundwater velocity and α is the heat diffusivity coefficient of water. More information can be found e.g. in Häfner et al. (1992) and Kolditz (1997).

The mesh for the numerical model consists of 501 nodes combining 500 line elements. The distance between the nodes is $\Delta x = 0.2 \, \text{m}$. The boundary conditions applied are as follows:

- Left border:

  - constant source term (liquid flow) with $Q = 3.0 \cdot 10^{-7} \, \text{m}^3 \text{s}^{-1}$
  - constant temperature with $T = 1 \, °\text{C}$

- Right border:

  - constant pressure with $P = 100 \, \text{kPa}$

- Initial conditions:

  - pressure with $P = 100 \, \text{kPa}$ for whole domain
  - temperature $T = 0 \, °\text{C}$ for whole domain

- Time step:

  - $\Delta t = 133 \, \text{s}$

With the given parameters, the Neumann criteria (4.2) results on Ne $= 0.5$ which guarantees the numerical stability of the diffusion part of the transport process. The Courant criteria, given by

$$\text{Cr} = \frac{v_x \cdot \Delta t}{\Delta x} \leq 1 \tag{4.4}$$

is equal to Cr $= 0.2$.

4.5.3   Input Files

After a very detailed description of the input files of the first example, we will highlight in the following only the new features used in the OGS input files. The file repository is www.opengeosys.org/tutorials/ces-i/e08. A brief overview of OGS keywords used in this tutorial can be found in Appendix B. Visit the OGS web-documentation http://www.opengeosys.org/help/documentation/ for more details.

4.5.3.1   GLI: Geometry

The geometry file simply contains two points demarking the model domain (fracture or column), basically for assigning boundary conditions. The two points are linked to a polyline which is used for data output along the $100\,m$ line model domain.

Listing 4.25 GLI input file

```
#POINTS
0  0  0  0  $NAME POINT0
1  100  0  0  $NAME POINT1
#POLYLINE
  $NAME
    PLY_0
  $POINTS
    0
    1
#STOP
```

4.5.3.2   MSH: Finite Element Mesh

The finite element mesh contains of 200 elements of half meter length formed by 201 grid nodes.

Listing 4.26 MSH input file

```
#FEM_MSH
$PCS_TYPE
HEAT_TRANSPORT
$NODES
201
0        0        0        0
1        0.5      0        0
2        1        0        0
...
198      99       0        0
199      99.5     0        0
200      100      0        0
$ELEMENTS
200
0        0        line     0        1
1        0        line     1        2
...
198      0        line     198      199
199      0        line     199      200
#STOP
```

4.5.3.3 PCS: Process Definition

This example has two processes `LIQUID_FLOW` and `HEAT_TRANSPORT` which are conducted sequentially within the time loop. The number of processes in the PCS file has consequences to almost all OGS objects and requires a corresponding number of keywords to define properties for all processes. The primary variable for liquid flow is pressure p, for heat transport is temperature T.

Listing 4.27 PCS input file

```
#PROCESS
 $PCS_TYPE
  LIQUID_FLOW
#PROCESS
 $PCS_TYPE
  HEAT_TRANSPORT
#STOP
```

4.5.3.4 NUM: Numerical Properties

As to the number of processes we need to define two NUM objects for each process. The subkeyword `$PCS_TYPE` indicates the corresponding process. The parameters for the linear equation solver are defined equally for both processes except of the error tolerance. Not the for heat transport process two Gauss points are prescribed for integration, the default value for line elements in one Gauss point.

Listing 4.28 NUM input file

```
#NUMERICS
 $PCS_TYPE
  LIQUID_FLOW
 $LINEAR_SOLVER
; method error_tolerance max_iterations theta precond storage
  2      2 1.e-016       1000           1.0   100     4
#NUMERICS
 $PCS_TYPE
  HEAT_TRANSPORT
 $LINEAR_SOLVER
; method error_tolerance max_iterations theta precond storage
  2      0 1.e-012       1000           1.0   100     4
 $ELE_GAUSS_POINTS
  2
#STOP
```

4.5.3.5 TIM: Time Discretization

The time stepping is identical for both processes: 2000 time steps with a length of 10^5 s are conducted. You should calculate the corresponding Courant number. Will the final time `$TIME_END` be reached?

Listing 4.29 TIM input file

```
#TIME_STEPPING
  $PCS_TYPE
    LIQUID_FLOW
  $TIME_STEPS
    2000  100000
  $TIME_END
    1.e9
  $TIME_START
    0
#TIME_STEPPING
  $PCS_TYPE
    HEAT_TRANSPORT
  $TIME_STEPS
    2000  100000
  $TIME_END
    1.e9
  $TIME_START
    0
#STOP
```

4.5.3.6   IC: Initial Conditions

Initial conditions are set for the primary variables of liquid flow (pressure) and heat transport (temperature). The number 1 at the end of the variable names indicates phase number 1. For multiphase flow processes or non-equilibrium thermal processes we have more than one phase. Initial conditions normally are valid for the entire model domain. The values are given as 10^5 Pa for liquid pressure and zero temperature. Note we are using relative values for temperature here ranging between 0 and 1.

Listing 4.30 IC input file

```
#INITIAL_CONDITION
  $PCS_TYPE
    LIQUID_FLOW
  $PRIMARY_VARIABLE
    PRESSURE1
  $GEO_TYPE
    DOMAIN
  $DIS_TYPE
    CONSTANT    100000
#INITIAL_CONDITION
  $PCS_TYPE
    HEAT_TRANSPORT
  $PRIMARY_VARIABLE
    TEMPERATURE1
  $GEO_TYPE
    DOMAIN
  $DIS_TYPE
    CONSTANT    0
#STOP
```

4.5.3.7   BC: Boundary Conditions

Dirichlet boundary conditions are given to both processes, a constant pressure of $10^5\,Pa$ in POINT1 (right boundary), a constant temperature of unity in POINT0 (left boundary). Note that non-specified boundaries for the finite element method are no flux conditions (Neumann zero condition).

Listing 4.31 BC input file

```
#BOUNDARY_CONDITION
 $PCS_TYPE
  LIQUID_FLOW
 $PRIMARY_VARIABLE
  PRESSURE1
 $GEO_TYPE
  POINT POINT1
 $DIS_TYPE
  CONSTANT    100000
#BOUNDARY_CONDITION
 $PCS_TYPE
  HEAT_TRANSPORT
 $PRIMARY_VARIABLE
  TEMPERATURE1
 $GEO_TYPE
  POINT POINT0
 $DIS_TYPE
  CONSTANT    1
#STOP
```

4.5.3.8   ST: Source/Sink Terms

For the flow process source term is added to the left boundary which is overwriting the no-flux condition. The source term (as of positive sign) is equal to a constant value of $10^{-6}\,m^3\,s^{-1}$. In order to calculate a corresponding Darcy velocity we need to know the cross-sectional area which is defined for the material group of the related finite element. We come back to this a little later with the MMP file definitions ...

Listing 4.32 ST input file

```
#SOURCE_TERM
 $PCS_TYPE
  LIQUID_FLOW
 $PRIMARY_VARIABLE
  PRESSURE1
 $GEO_TYPE
  POINT POINT0
 $DIS_TYPE
  CONSTANT_NEUMANN 1E-6
#STOP
```

4.5.3.9   MFP: Fluid Properties

Typical fluid properties of water are used (which p-T conditions?) (see Table 4.8).

Listing 4.33 MFP input file

```
#FLUID_PROPERTIES
 $FLUID_TYPE
  LIQUID
 $DENSITY
  1 1000
  $VISCOSITY
  1 0.001
 $SPECIFIC_HEAT_CAPACITY
  1 4000
 $HEAT_CONDUCTIVITY
  1 0.6
#STOP
```

4.5.3.10   MSP: Solid Properties

Mechanical (density) and thermophysical properties of the solid phase of the porous medium are defined (see Table 4.8). All parameter model types are 1 indicating constant material properties (linear models).

Listing 4.34 MSP input file

```
#SOLID_PROPERTIES
 $DENSITY
 1 2850
 $THERMAL
  EXPANSION
   1.0e-5
  CAPACITY
   1 6000.
  CONDUCTIVITY
   1 5
#STOP
```

4.5.3.11   MMP: Medium Properties

The parameters for the porous medium are given in Table 4.8 as well. All parameter model types are 1 indicating constant material properties (linear models). We need some more explanation for the following subkeywords:

- $GEOMETRY_DIMENSION: This is the geometric dimension of this material group:
  - This geometric dimension is necessary to combine different-dimensions element in 3-D space (see $GEOMETRY_AREA)
  - The geometric material group dimension is also important for correct tensor calculations (see $PERMEABILITY_TENSOR)

- $GEOMETRY_AREA:

This is a geometric property of the related finite element. Here we have simply 1-D elements, therefore, the geometry dimensions describes the cross-section area of the element, i.e. 1 m². As of the 1 square meter cross-section area of the element, the corresponding Darcy velocity resulting from the source term is equal to

$$\mathbf{q} = \frac{Q}{A} = \frac{10^{-6}\,\mathrm{m^3 s^{-1}}}{1\,\mathrm{m^2}} = 10^{-6}\,\mathrm{ms^{-1}} \tag{4.5}$$

This is something we need to check in the results later on ...

- $PERMEABILITY_TENSOR: The tensor type is indicated (isotropic) followed by the values, here just one value as of isotropy.
- $HEAT_DISPERSION: Heat dispersion has two numbers, heat dispersion length in longitudinal α_L and transverse directions α_T.

Listing 4.35 MMP input file

```
#MEDIUM_PROPERTIES
 $GEOMETRY_DIMENSION
  1
 $GEOMETRY_AREA
  1.0
 $POROSITY
  1 1
 $STORAGE
  1 0.0
 $TORTUOSITY
  1 1.0
 $PERMEABILITY_TENSOR
  ISOTROPIC 1.00000e-11
 $HEAT_DISPERSION
  1  0.000000e+000 0.000000e+000
#STOP
```

4.5.3.12 OUT: Output Parameters

Finally we define output data. For multiple processes we can output multiple primary variables at geometries. Here we output pressure, temperature, and velocity in x direction at point POINT1 (right boundary). The output data type is for TECPLOT and results are written every time step to the output file.

Listing 4.36 OUT input file

```
#OUTPUT
 $NOD_VALUES
  PRESSURE1
  TEMPERATURE1
  VELOCITY_X1
 $GEO_TYPE
  POINT POINT1
 $DAT_TYPE
```

```
TECPLOT
$TIM_TYPE
  STEPS 1
#STOP
```

4.5.3.13   File Repository

The file repository is www.opengeosys.org/tutorials/ces-i/e08

4.5.4   Results

In Fig. 4.14 a comparison of the analytical and numerical solutions is plotted. The figure shows the temperature breakthrough curve at the end of the fracture at $x = 100\,m$. The numerical results show acceptable agreement with the analytical solution. In a further step, the diffusion part of the heat transport process was avoided by minimizing the thermal conductivity of the fluid. Figure 4.15 shows the breakthrough curve for only advective heat transport.

Figure 4.16 shows the temperature distribution (profile) along the fracture after about 14 h.

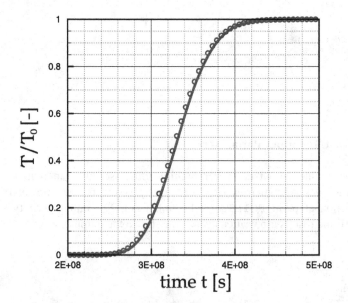

Fig. 4.14 Temperature breakthrough curve at the point $x = 100\,m$

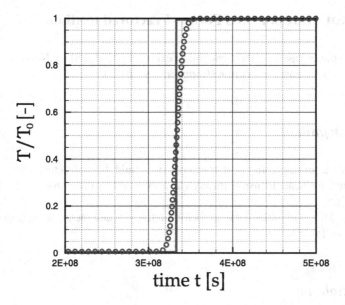

Fig. 4.15 Temperature breakthrough curve when diffusion is neglected (shows numerical diffusion)

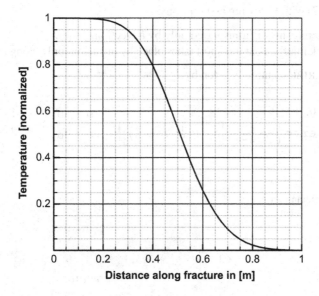

Fig. 4.16 Temperature profile along the fracture after $t = 50,000\,s$

4.6 Heat Transport in a Porous-Fractured medium

This example is based on the benchmark exercise introduced by Norbert Böttcher to the OGS benchmark collection (Böttcher et al. 2012).

4.6.1 Definition

Based on the example for heat transport in a fluid filled fracture, this problem is extended by heat diffusion through a rock matrix orthogonal to the fracture (Fig. 4.17).

The model and material parameters for the fracture and rock matrix, respectively, are given in Table 4.9.

4.6.2 Solution

For this problem an analytical solution was derived by LAUWERIER (1955) (see Kolditz 1997) with following assumptions:

- in the fracture, heat is transported only by advection,
- in the rock matrix, heat transport takes place by diffusion (only along the z-axis).

The LAUWERIER solution is given by

$$
T_D = \begin{cases} 0, & t_D < x_D \\ \mathtt{erfc}\left\{ \frac{\beta}{\sqrt{\alpha(t_D - x_D)}} \left[x_D + \frac{1}{2\beta}\left(z_D - \frac{1}{2}\right)\right]\right\}, & t_D > x_D \end{cases}, z_D \geq \frac{1}{2} \qquad (4.6)
$$

Fig. 4.17 Heat transport in a fracture-matrix system

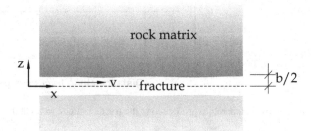

Table 4.9 Model parameters for the LAUWERIER-problem

Symbol	Parameter	Value	Unit
Spatial discretisation			
L	Fracture length	50	m
W	Matrix width	63.25	m
Δx	Step size X	2	m
Δz	Step size Z	0.1265	m
$b/2$	Half of fracture width	$1.0 \cdot 10^{-3}$	m
v_x	Groundwater velocity	$1.0 \cdot 10^{-4}$	$\mathrm{m\,s^{-1}}$
Temporal discretisation			
Δt	Time step length	$2.0 \cdot 10^5$	s
	Number of time steps	2500	
	Total time	$5.0 \cdot 10^8$	s
Material properties—solid			
λ	Thermal conductivity	1	$\mathrm{W\,m^{-1}\,K^{-1}}$
c	Heat capacity	1000	$\mathrm{J\,kg^{-1}\,K^{-1}}$
ρ	Density	2500	$\mathrm{kg\,m^{-3}}$
Material properties—fluid			
c	Heat capacity	4000	$\mathrm{J\,kg^{-1}\,K^{-1}}$
ρ	Density	1000	$\mathrm{kg\,m^{-3}}$

Fig. 4.18 Alignment of the grid for the numerical model

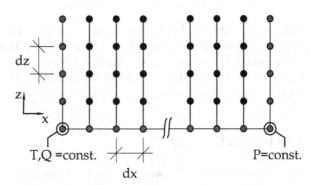

with the following dimensionless parameters:

$$t_D = \frac{v_x}{b}t, \quad x_D = \frac{x}{b}, \quad z_D = \frac{z}{b}, \quad \alpha = \frac{\lambda^s}{c^s \rho^s} \frac{1}{b v_x}, \quad \beta = \frac{\lambda^s}{c^l \rho^l} \frac{1}{b v_x} \qquad (4.7)$$

where b is the fracture width, λ is the thermal conductivity, c is the heat capacity, ρ is the density and the suffixes s and f denote the solid (rock) and liquid (water) phases, respectively.

The numerical LAUWERIER model is formed as a coupling of advective 1D heat transport in x-direction and diffusive 1D heat transport in z-direction. This means, that nodes in the rock matrix are not influenced by their left or right neighbors. The matrix elements are connected to the fracture elements orthogonally. Figure 4.18 shows a schematical description of the model setup. Because of the symmetry, the numerical model calculates just the domain above the x-axis.

Fig. 4.19 Positions of observation points for temperature breakthrough curves

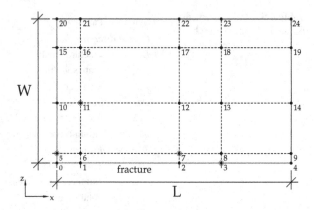

Figure 4.19 shows the positions of observation points which were chosen to evaluate the numerical model in comparison with analytical solutions.

4.6.3   Input files

After a very detailed description of the input files of the first example, we will highlight in the following only the new features used in the OGS input files. The file repository is www.opengeosys.org/tutorials/ces-i/e09. A brief overview of OGS keywords used in this tutorial can be found in Appendix B. Visit the OGS web-documentation http://www.opengeosys.org/help/documentation/ for more details.

4.6.3.1   GLI: Geometry

The geometry for the Lauwerier example is more complex as we combine 1-D elements into a 2-D model. It looks like a "brush", see Fig. 4.18 showing the alignment of the grid for the numerical model.

- #POINTS: The points POINT0-POINT4 represent the fracture with a length of 50 m The points POINT5-POINT9 represent a parallel line to the fracture at a distance of 0.253 m from the fracture and so on up to a distance of 63.25 m from the fracture line. Some points are used simply for data output, see Fig. 4.19 depicting the positions of observation points for temperature breakthrough curves.
- #POLYLINE: The polylines describe the fracture (named FRACTURE) as well as horizontal (parallel to the fracture—named H) and vertical lines (orthogonal to the fracture—named V)

Listing 4.37 GLI input file

```
#POINTS
0        0        0        0 $NAME POINT0
1        2        0        0 $NAME POINT1
2        10       0        0 $NAME POINT2
3        20       0        0 $NAME POINT3
4        50       0        0 $NAME POINT4
5        0        0        0.253 $NAME POINT5
6        2        0        0.253 $NAME POINT6
7        10       0        0.253 $NAME POINT7
8        20       0        0.253 $NAME POINT8
9        50       0        0.253 $NAME POINT9
...
20       0        0        63.25 $NAME POINT20
21       2        0        63.25 $NAME POINT21
22       10       0        63.25 $NAME POINT22
23       20       0        63.25 $NAME POINT23
24       50       0        63.25 $NAME POINT24
#POLYLINE
 $NAME
  FRACTURE
 $POINTS
  0
  4
...
#POLYLINE
 $NAME
  H1
 $POINTS
  5
  9
...
#POLYLINE
 $NAME
  V1
 $POINTS
  1
  21
#STOP
```

4.6.3.2   MSH: Finite Element Mesh

- $NODES: The mesh nodes are arranged line-wise beginning with the fracture. A fine resolution is required for the Lauwerier problem to achieve accurate solutions resulting in a number of 13026 grid nodes—an exercise which is not recommended doing manually ;-).
- $ELEMENTS: We have three types of elements:

    - fracture elements: forming the fracture (material group 0)
    - matrix boundary elements: building the matrix at the model boundary (material group 2)
    - matrix inner elements: building the matrix in the inner the model domain (material group 1)

This is due to suit the geometric combination of 1-D elements in 2-D, e.g. the matrix boundary elements have only half of the contact line to the fracture than the inner matrix elements. We come back to this specifics when we explain the medium material properties (MMP objects)

Listing 4.38   MSH input file

```
#FEM_MSH
$NODES
13026
0        0.00      0.00      0.00 ; fracture nodes
1        2.00      0.00      0.00 ; fracture nodes
...
24       48.00     0.00      0.00 ; fracture nodes
25       50.00     0.00      0.00 ; fracture nodes
...
13024    48.00     0.00      63.25 ; matrix nodes
13025    50.00     0.00      63.25 ; matrix nodes

$ELEMENTS
13025
0        0        -1       line    0       1 ; fracture elements
1        0        -1       line    1       2
...
23       0        -1       line    23      24
24       0        -1       line    24      25 ; fracture elements
25       2        -1       line    0       26 ; matrix boundary elements
26       2        -1       line    26      52
...
523      2        -1       line    12948   12974
524      2        -1       line    12974   13000 ; matrix boundary elements
525      1        -1       line    1       27 ; matrix inner elements
526      1        -1       line    27      53
...
12523    1        -1       line    12972   12998
12524    1        -1       line    12998   13024 ; matrix inner elements
12525    2        -1       line    25      51 ; matrix boundary elements
12526    2        -1       line    51      77
...
13023    2        -1       line    12973   12999
13024    2        -1       line    12999   13025
#STOP
```

4.6.3.3   PCS: Process Definition

We define two processes for liquid flow and heat transport. By using the subkeyword $TIM_TYPE with the option STEADY, we indicate that the hydraulic part of or example is a steady-state problem (i.e. it is not depending on time). This means, that no matter how we define the temporal discretization of the hydraulic problem, the computation occurs only once, saving a lot of computing time.

Listing 4.39   PCS input file

```
#PROCESS
 $PCS_TYPE
  LIQUID_FLOW
 $TIM_TYPE
  STEADY
```

```
#PROCESS
 $PCS_TYPE
  HEAT_TRANSPORT
#STOP
```

4.6.3.4   NUM: Numerical Properties

We use different error calculation methods for flow and transport processes.

Listing 4.40 NUM input file

```
#NUMERICS
 $PCS_TYPE
  LIQUID_FLOW
 $LINEAR_SOLVER
; method error_tolerance max_iterations theta precond storage
  2       2 1.e-016        1000           1.0   100     4
#NUMERICS
 $PCS_TYPE
  HEAT_TRANSPORT
 $LINEAR_SOLVER
; method error_tolerance max_iterations theta precond storage
  2       0 1.e-012        1000           1.0   100     4
 $ELE_GAUSS_POINTS
  2
#STOP
```

4.6.3.5   TIM: Time Discretization

We use typical input for time discretization according the Table 4.9.

Listing 4.41 TIM input file

```
#TIME_STEPPING
...
```

4.6.3.6   IC: Initial Conditions

Initial conditions of both processes are set for the entire domain.

Listing 4.42 IC input file

```
#INITIAL_CONDITION
 $PCS_TYPE
  LIQUID_FLOW
 $PRIMARY_VARIABLE
  PRESSURE1
 $GEO_TYPE
  DOMAIN
 $DIS_TYPE
  CONSTANT    100000
```

```
#INITIAL_CONDITION
 $PCS_TYPE
  HEAT_TRANSPORT
 $PRIMARY_VARIABLE
  TEMPERATURE1
 $GEO_TYPE
  DOMAIN
 $DIS_TYPE
  CONSTANT    0
#STOP
```

4.6.3.7   BC: Boundary Conditions

Two Dirichlet boundary conditions are assigned at the fracture outlet (constant pressure) and at the fracture inlet (constant temperature).

Listing 4.43 BC input file

```
#BOUNDARY_CONDITION
 $PCS_TYPE
  LIQUID_FLOW
 $PRIMARY_VARIABLE
  PRESSURE1
 $GEO_TYPE
  POINT POINT4
 $DIS_TYPE
  CONSTANT    100000
#BOUNDARY_CONDITION
 $PCS_TYPE
  HEAT_TRANSPORT
 $PRIMARY_VARIABLE
  TEMPERATURE1
 $GEO_TYPE
  POINT POINT0
 $DIS_TYPE
  CONSTANT    1
#STOP
```

4.6.3.8   ST: Source/Sink Terms

A fluid source term is given to the left side of the fracture (POINT0). The ST value in combination with the cross-section area of the fracture elements (see first MMP object) gives the following flow velocity

$$\mathbf{q} = \mathbf{v}n = \frac{Q}{A} = \frac{10^{-7}\,\text{m}^3\,\text{s}^{-1}}{10^{-3}\,\text{m}^2} = 10^{-4}\,\text{m}\,\text{s}^{-1} \tag{4.8}$$

Listing 4.44 ST input file

```
#SOURCE_TERM
 $PCS_TYPE
  LIQUID_FLOW
```

```
$PRIMARY_VARIABLE
  PRESSURE1
$GEO_TYPE
  POINT POINT0
$DIS_TYPE
  CONSTANT 1e-7
#STOP
```

4.6.3.9 MFP: Fluid Properties

We use typical input of fluid properties according the Table 4.9.

Listing 4.45 MFP input file

```
#FLUID_PROPERTIES
...
```

4.6.3.10 MSP: Solid Properties

As we need 3 material groups to distinguish between the geometric properties (see MSH file description) we also have to define 3 solid properties which are identical.

Listing 4.46 MSP input file

```
#SOLID_PROPERTIES
 $DENSITY
  1 2500
 $THERMAL
  EXPANSION:
   1 0
  CAPACITY:
   1 1000
  CONDUCTIVITY:
   1 1
#SOLID_PROPERTIES
...
#SOLID_PROPERTIES
...
#STOP
```

4.6.3.11 MMP: Medium Properties

- fracture elements: forming the fracture (material group 0): the cross-section area of the 1-D fracture elements is 10^{-3} m^2, e.g. 10^{-3} m $\times$ 1 m, fracture porosity is 1 and the fracture has a permeability value, no heat dispersion just diffusivity.
- matrix inner elements: building the matrix in the inner the model domain (material group 1): the contact area of the 1-D fracture elements is 2 m^2, i.e. 2 m (length) $\times$ 1 m (width), fracture porosity is 0 and the fracture has a zero permeability value, no heat dispersion just diffusivity.

- matrix boundary elements: building the matrix at the model boundary (material group 2): The only difference of the matrix boundary element is half of the contact area, i.e. 1 m (length) × 1 m (width).

Listing 4.47   MMP input file

```
#MEDIUM_PROPERTIES
 $GEOMETRY_DIMENSION
  1
 $GEOMETRY_AREA
  1.0E-3
 $POROSITY
  1  1
 $STORAGE
  1  0.0
 $TORTUOSITY
  1  1.000000e+000
 $PERMEABILITY_TENSOR
  ISOTROPIC   1.0e-15
 $HEAT_DISPERSION
  1  0.000000e+000 0.000000e+000
#MEDIUM_PROPERTIES
 $GEOMETRY_DIMENSION
  1
 $GEOMETRY_AREA
  2.0E+0
 $POROSITY
  1  0
 $STORAGE
  1  0.0
 $TORTUOSITY
  1  1.000000e+000
 $PERMEABILITY_TENSOR
  ISOTROPIC   0
 $HEAT_DISPERSION
  1  0.000000e+000 0.000000e+000
#MEDIUM_PROPERTIES
 $GEOMETRY_DIMENSION
  1
 $GEOMETRY_AREA
  1.0E+0
  ...
#STOP
```

4.6.3.12   OUT: Output Parameters

Data output is given along the fracture polyline FRACTURE and in one observation POINT1 every ten time steps.

Listing 4.48   OUT input file

```
#OUTPUT
 $NOD_VALUES
  TEMPERATURE1
 $GEO_TYPE
  POINT POINT1
 $TIM_TYPE
  STEPS 10
#OUTPUT
 $NOD_VALUES
```

Fig. 4.20 Temperature distribution orthogonal to the fracture at $x = 0$ m at three different times

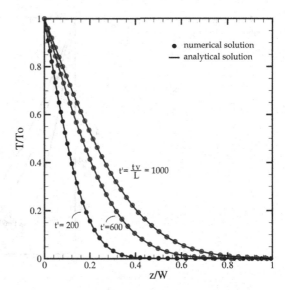

```
TEMPERATURE1
$GEO_TYPE
  POLYLINE FRACTURE
$TIM_TYPE
  STEPS 10
#STOP
```

4.6.3.13   File Repository

The file repository is www.opengeosys.org/tutorials/ces-i/e09

4.6.4   Results

The quality of the numerical results can be shown by temperature distribution curves for several times in the rock matrix. Figure 4.20 shows the temperature profiles for $x = 0$ m at three moments t'. The numerical solution is in very good agreement with the analytical results. Temperature profiles along the fracture at $z = 0$ m are plotted in Figs. 4.21 and 4.22.

For long simulation times ($t' = 1000; t' = 600$) both solutions fit very well together. For short simulation times, the numerical solution differs slightly from the analytical results. This discrepancy for short simulation times can be examined in Fig. 4.23, where temperature breakthrough curves for certain points (see Fig. 4.19) are plotted.

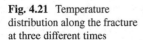 **Fig. 4.21** Temperature distribution along the fracture at three different times

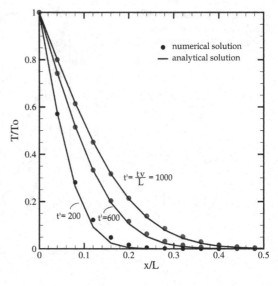

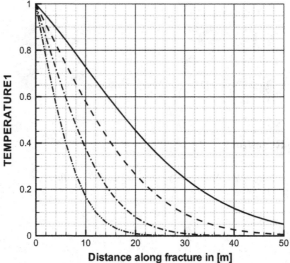

Fig. 4.22 Temperature distribution along the fracture at different times in real times: check dimensionless parameters

4.7 Heat Convection in a Porous Medium: The Elder problem

This example is based on the benchmark exercise introduced by Norihiro Watanabe to the OGS benchmark collection. This exercise is dealing with thermal convection in porous media, which is an example of nonlinear flow problems. Thermal

Fig. 4.23 Temperature breakthrough curves at certain points in the rock matrix

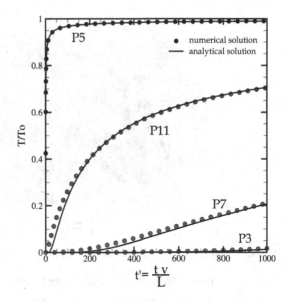

convection phenomena are present in geological as well hydrogeological systems when sufficient large temperature changes occur.

Thermal convection phenomena can occur for a large range of hydrodynamic conditions which are characterized by corresponding characteristic numbers (Elder 1977)

- very small Reynolds number (slow viscous flow),
- very large Prandtl number (heat transfer independent of inertial effects),
- large Peclet numbers (advection dominant in forced convection),
- very large Rayleigh number (free convection strong and turbulent).

4.7.1   Definition

The definition of the thermal Elder problem is depicted in Fig. 4.24. We consider a vertical (half) domain of 300 m length and 150 m height. The boundary conditions are defined as follows:

- Fixed pressure $p = 0$ Pa at the upper left corner,
- Fixed temperature $T = 1\,°C$ at a part of the bottom,
- Fixed temperature $T = 0\,°C$ at the top,
- Everywhere else we have no flow boundaries for both fluid and heat fluxes.

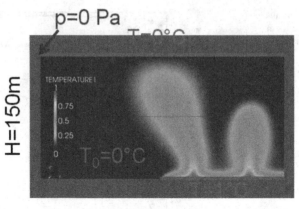

Fig. 4.24 Definition of the thermal Elder problem

Table 4.10 Model parameters for the ELDER-problem

Symbol	Parameter	Value	Unit
Fluid properties			
ρ_0^f	Reference density	1000	$\mathrm{kg\,m^{-3}}$
μ^f	Viscosity	10^{-3}	$\mathrm{Pa\,s}$
c^f	Specific heat capacity	4200	$\mathrm{J\,kg^{-1}\,K^{-1}}$
λ^f	Thermal conductivity	0.65	$\mathrm{W\,m^{-1}\,K^{-1}}$
Solid properties			
ρ_0^s	Reference density	0	$\mathrm{kg\,m^{-3}}$
c^s	Specific heat capacity	850	$\mathrm{J\,kg^{-1}\,K^{-1}}$
λ^f	Thermal conductivity	1.591444	$\mathrm{W\,m^{-1}\,K^{-1}}$
Medium properties			
n	Porosity	0.1	$\mathrm{m^3\,m^{-3}}$
τ	Tortuosity	1	$\mathrm{m^3\,m^{-3}}$
k	Permeability	$4.84404 \cdot 10^{-13}$	$\mathrm{m^2}$
S_s	Specific storage	0	$\mathrm{Pa^{-1}}$
$\alpha_{L|T}$	Heat dispersion length	0	m

The main difference to the previous exercises is that the fluid density is not constant but depending on temperature (Table 4.10):

$$\rho^f(T) = \rho_0^f \left(1 - \alpha_T\, T\right) \tag{4.9}$$

$$\rho^f(T) = 1000 \frac{\mathrm{kg}}{\mathrm{m^3}} \left(1 - 0.2\, T\right) \tag{4.10}$$

4.7.2 Input files

After a very detailed description of the input files of the first example, we will highlight in the following only the new features used in the OGS input files. The file repository is www.opengeosys.org/tutorials/ces-i/e10. A brief overview of OGS keywords used in this tutorial can be found in Appendix B. Visit the OGS web-documentation http://www.opengeosys.org/help/documentation/ for more details.

4.7.2.1 GLI: Geometry

Points and polylines are defined for describing the model domain as well as boundary conditions according to the benchmark definition.

* #POINTS: for corner points or the model domain as well as mid-point of top and bottom lines,
* #POLYLINE: for boundary conditions on top and half bottom.

Listing 4.49 GLI input file

```
#POINTS
0    0 0     0 $NAME POINT0
1    0 0 -150 $NAME POINT1
2 300 0 -150 $NAME POINT2
3 300 0    0 $NAME POINT3
4 150 0    0 $NAME POINT4
5 150 0 -150 $NAME POINT5
#POLYLINE
 $NAME
  TOP
 $TYPE
  2
 $POINTS
  3
  0
#POLYLINE
 $NAME
  BOTTOM_HALF_RIGHT
 $TYPE
  2
 $POINTS
  5
  2
#STOP
```

4.7.2.2 MSH: Finite Element Mesh

For mesh construction we use Gmsh software (Geuzaine and Remacle 2009) as we want to be flexible in mesh resolution. Moreover the number of mesh nodes and elements becomes too large for manual construction. We use a structured mesh consisting of linear quad elements.

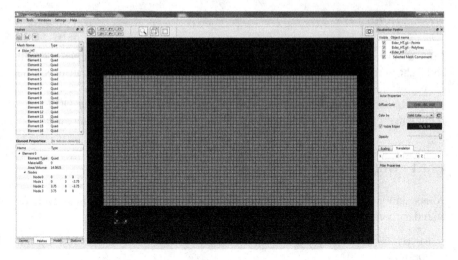

Fig. 4.25 Display of Elder mesh with ogs[6] DataExplorer

- $NODES: along vertical lines,
- $ELEMENTS: as horizontal stripes.

Listing 4.50 MSH input file

```
#FEM_MSH
 $NODES
  3321
0 0 0 0
1 0 0 -3.75
...
3319 296.25 0 -150
3320 300 0 -150
 $ELEMENTS
  3200
0 0 quad 0 1 2 3
1 0 quad 3 2 4 5
...
3198 0 quad 3237 3318 3319 3238
3199 0 quad 3238 3319 3320 3239
#STOP
```

The mesh can be also displayed using the ogs[6] DataExplorer (Fig. 4.25).

4.7.2.3   PCS: Process Definition

Two processes are defined, liquid flow and heat transport. Note, that there is no difference so far to linear heat transport processes in the previous exercises. The process coupling is defined in the numerical properties (i.e. NUM files).

Listing 4.51 PCS input file

```
#PROCESS
 $PCS_TYPE
  LIQUID_FLOW
#PROCESS
 $PCS_TYPE
  HEAT_TRANSPORT
#STOP
```

4.7.2.4 NUM: Numerical Properties

In addition to the linear solver controls, the numerical properties describe also the non-linear solver procedures. In this example, we need the non-linear solver controls for heat transport because the fluid density depends on temperature.

Furthermore, we need to provide a coupling control since we solve liquid flow and heat transport separately. Coupling of the two processes can be achieved by iteratively solving them until the solutions get convergence. The coupling control can be configured by the following keywords,

- $OVERALL_COUPLING: The ranges (min,max) of iterations for overall couplings of several processes are defined (Table 4.11).
- $COUPLING_CONTROL: The error method and the error tolerances are given for individual processes (Table 4.12).

Listing 4.52 NUM input file

```
$OVERALL_COUPLING
 2 25 ; min_iter | max_iter
#NUMERICS
 $PCS_TYPE
  LIQUID_FLOW
 $LINEAR_SOLVER
  2     2 1.e-016      1000        1.0  100      4
 $ELE_GAUSS_POINTS
  3
 $COUPLING_CONTROL
   LMAX 10 ; error method | tolerance
#NUMERICS
 $PCS_TYPE
  HEAT_TRANSPORT
 $LINEAR_SOLVER
  2     0 1.e-012      1000        1.0  100      4
 $ELE_GAUSS_POINTS
  3
 $NON_LINEAR_ITERATIONS
;type -- error_method -- max_iterations -- relaxation -- tolerance(s)
  PICARD LMAX 25 0.0 1e-3
 $COUPLING_CONTROL
   LMAX 1.e-3 ; error method -- tolerances
#STOP
```

Table 4.11 Keyword:
$OVERALL_COUPLING

Parameter	Meaning	Values
min_iter	Minimum number of overall iterations	2
max_iter	Maximum number of overall iterations	25

Table 4.12 Keyword:
$COUPLING_CONTROL

Parameter	Meaning	Values
Method	Error method	LMAX
Tolerance	Error tolerance	Process dependent

4.7.2.5   TIM: Time Discretization

We use a fixed time stepping for both processes.

Listing 4.53 TIM input file

```
#TIME_STEPPING
  $PCS_TYPE
   LIQUID_FLOW
  $TIME_START
    0.0
  $TIME_END
    126144000 ; 4 years
  $TIME_STEPS
    48 2628000 ; 1 month
#TIME_STEPPING
  $PCS_TYPE
    HEAT_TRANSPORT
...
#STOP
```

4.7.2.6   IC: Initial Conditions

Initial conditions of both processes are set for the entire domain.

Listing 4.54 IC input file

```
#INITIAL_CONDITION
  $PCS_TYPE
   LIQUID_FLOW
  $PRIMARY_VARIABLE
   PRESSURE1
  $GEO_TYPE
   DOMAIN
  $DIS_TYPE
   CONSTANT    0
#INITIAL_CONDITION
  $PCS_TYPE
   HEAT_TRANSPORT
  $PRIMARY_VARIABLE
   TEMPERATURE1
  $GEO_TYPE
   DOMAIN
  $DIS_TYPE
   CONSTANT    0
#STOP
```

4.7.2.7   BC: Boundary Conditions

Constant Dirichlet boundary conditions are assigned according to the benchmark definition (Fig. 4.24).

Listing 4.55 BC input file

```
#BOUNDARY_CONDITION
 $PCS_TYPE
  LIQUID_FLOW
 $PRIMARY_VARIABLE
  PRESSURE1
 $GEO_TYPE
  POINT POINT0
 $DIS_TYPE
  CONSTANT   0
#BOUNDARY_CONDITION
 $PCS_TYPE
  HEAT_TRANSPORT
 $PRIMARY_VARIABLE
  TEMPERATURE1
 $GEO_TYPE
  POLYLINE BOTTOM_HALF_RIGHT
 $DIS_TYPE
  CONSTANT    1
#BOUNDARY_CONDITION
 $PCS_TYPE
  HEAT_TRANSPORT
 $PRIMARY_VARIABLE
  TEMPERATURE1
 $GEO_TYPE
  POLYLINE TOP
 $DIS_TYPE
  CONSTANT   0
#STOP
```

4.7.2.8   ST: Source/Sink Terms

No source terms are applied.

4.7.2.9   MFP: Fluid Properties

The fluid density is a linear function of temperature (4.10). An overview of existing fluid density models can be found in Sect. B.1 in Appendix B.

Listing 4.56 MFP input file

```
#FLUID_PROPERTIES
 $DENSITY
  4 1000 0 -0.2
 $VISCOSITY
  1 0.001
 $SPECIFIC_HEAT_CAPACITY
  1 4200.0
 $HEAT_CONDUCTIVITY
  1 0.65
```

```
#STOP

Remove from input file:
 $FLUID_TYPE
  LIQUID
 $FLUID_NAME
  WATER
```

4.7.2.10   MSP: Solid Properties

Constant solid properties are used (Table 4.10).

Listing 4.57 MSP input file

```
#SOLID_PROPERTIES
 $DENSITY
  1 0
  $THERMAL
  EXPANSION
  1.000000000000e-005
  CAPACITY
  1 8.500000000000e+002
  CONDUCTIVITY
  1 1.591444000000e+000
#STOP
```

4.7.2.11   MMP: Medium Properties

According to the benchmark definition (Table 4.10) the following medium properties are used. Storativity is set to zero for steady state flow conditions. Heat dispersion is neglect as to the benchmark definition.

Listing 4.58 MMP input file

```
#MEDIUM_PROPERTIES
 $GEOMETRY_DIMENSION
  2
 $POROSITY
  1  0.1
 $TORTUOSITY
  1  1.000000e+000
 $PERMEABILITY_TENSOR
  ISOTROPIC  4.84404E-13
 $STORAGE
  1 0.0
 $HEAT_DISPERSION
  1  0  0
#STOP

Remove from input file and test:
 $MASS_DISPERSION
  1  0  0
```

4.7.2.12   OUT: Output Parameters

Data output is given for the entire domain every for time step. The output format used is VTK which can be displayed by ParaView or the OGS DataExplorer.

Listing 4.59 OUT input file

```
#OUTPUT
 $NOD_VALUES
  PRESSURE1
  TEMPERATURE1
$ELE_VALUES
  VELOCITY1_X
  VELOCITY1_Y
 $GEO_TYPE
  DOMAIN
 $DAT_TYPE
  PVD
 $TIM_TYPE
  STEPS 1
#STOP
```

4.7.2.13   File Repository

The file repository is www.opengeosys.org/tutorials/ces-i/e10.

4.7.3   Results

Simulation results from the last time step ($t= 126144000\,$s = 4 years) are displayed in two different ways in Fig. 4.26 using ParaView and Fig. 4.27 using the OGS DataExplorer. Thermal buoyancy processes result in upwelling convection cells shows by the curved temperature isolines.

4.7.4   Exercises

1. Use GINA for set-up geometries and finite element meshes.
2. Try different mesh densities.
3. Temperature dependent viscosity.
4. Effect of heat dispersion.

Fig. 4.26 Elder problem displayed with ParaView

Fig. 4.27 Elder problem displayed with OGS DataExplorer

Chapter 5
Introduction to Geothermal Case Studies

This section is an appetizer to continue working with the OGS tutorials—after a rather theoretical introduction. We show some of the OGS applications in advanced geothermal reservoir modeling. You also may look into Huenges (2011) for further geothermal modeling applications ...

5.1 Bavarian Molasse

We literally cite from Cacace et al. (2013) "Fluid flow in low-permeable carbonate rocks depends on the density of fractures, their inter-connectivity and on the formation of fault damage zones. The present day stress field influences the aperture hence the transmissivity of fractures whereas paleo stress fields are responsible for the formation of faults and fractures. In low-permeable reservoir rocks, fault zones belong to the major targets. Before drilling, an estimate for reservoir productivity of wells drilled into the damage zone of faults is therefore required. Due to limitations in available data, a characterization of such reservoirs usually relies on the use of numerical techniques. The requirements of these mathematical models encompass a full integration of the actual fault geometry, comprising the dimension of the fault damage zone and of the fault core, and the individual population with properties of fault zones in the hanging and foot wall and the host rock. The paper presents both the technical approach to develop such a model and the property definition of heterogeneous fault zones and host rock with respect to the current stress field. The case study describes a deep geothermal reservoir in the western central Molasse Basin in southern Bavaria, Germany. Results from numerical simulations indicate that the well productivity can be enhanced along compressional fault zones if the inter-connectivity of fractures is lateral caused by crossing synthetic

© Springer International Publishing Switzerland 2016
N. Böttcher et al., *Geoenergy Modeling I*, SpringerBriefs in Energy,
DOI 10.1007/978-3-319-31335-1_5

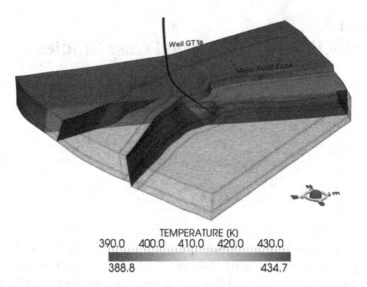

Fig. 5.1 Temperature boundary conditions applied during the simulation (Cacace et al. 2013)

and antithetic fractures. The model allows a deeper understanding of production
tests and reservoir properties of faulted rocks." Figure 5.1 depicts the temperature
boundary conditions posed on the reservoir for numerical analysis.

5.2   Urach Spa

Urach Spa was one of the most important scientific geothermal pilot projects in
Germany even though it failed but it provided deep insight into the complexity of
deep geothermal systems (Tenzer et al. 2010). Watanabe et al. (2010) developed
a numerical THM model for the Urach Spa location. The proposed boreholes
(U3 and U4) of the dipole flow circulation system (i.e. a "doublet") are located
400 m apart. Parameters relevant to reservoir fluid flow and heat transport that were
used in the model were based on the results of previous studies. The hydraulically
active areas allow the reservoir to be represented geometrically as a cuboid which is
300 m high, 300 m wide and 800 m long. Figure 5.2 shows the simulated flow field
and temperature distribution during exploiting the geothermal reservoir. The Urach
Spa project was also investigated in scientific visualization exercise for gaining
additional knowledge from complex simulations studies (Zehner et al. 2010).

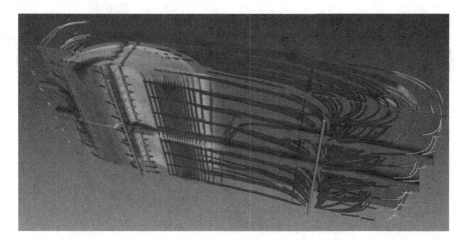

Fig. 5.2 Numerical simulation and visualization of the geothermal Urach Spa research location

Fig. 5.3 Geothermal
reservoir simulation

5.3 Gross Schönebeck

Figure 5.3 depicts the application area: **Geothermal energy,** which is one of the
alternative future energy resources under consideration. So-called shallow and deep
geothermal systems are distinguished. Shallow systems are already commercially
used e.g. for heating purposes. Deep geothermal reservoirs can be used for electric

power production as high temperatures up to 200 °C can be produced. THM/C modeling is required to design these geothermal power plants, e.g. in order to optimize production efficiency and reservoir lifetime. The significant cooling of the reservoir due to fluid reinjection gives rise to thermo-mechanical effects which need to be controlled in order to avoid reservoir damage (Watanabe et al. 2012). The geothermal study of the Gross Schönebeck research site was also selected as a scientific visualization study for dealing with complex and big geoscientific data sets.

Appendix A
Symbols

© Springer International Publishing Switzerland 2016

N. Böttcher et al., *Geoenergy Modeling I*, SpringerBriefs in Energy,

DOI 10.1007/978-3-319-31335-1

Table A.1 Table of symbols

Symbol	Parameter	Unit
Latin symbols		
$\mathbf{A}$	Global system matrix	
a	Heat transfer coefficient	$\mathrm{W\,K^{-1}\,m^{-2}}$
$\mathbf{b}$	Right-hand-side vector	
c	Specific heat capacity	$\mathrm{J\,kg^{-1}\,K^{-1}}$
Cr	Courant number, criteria	
$\mathbf{D}$	Diagonal matrix	
e	Specific energy	$\mathrm{J\,kg^{-1}}$
e_k	Iteration error	
$\mathbf{g}$	Gravity acceleration vector	$\mathrm{m\,s^{-1}}$
h	Specific enthalpy	$\mathrm{J\,kg^{-1}}$
$\mathbf{j}_{\mathrm{adv}}$	Advective heat flux	$\mathrm{W\,m^{-2}}$
$\mathbf{j}_{\mathrm{diff}}$	Diffusive heat flux	$\mathrm{W\,m^{-2}}$
$\mathbf{j}_{\mathrm{disp}}$	Dispersive heat flux	$\mathrm{W\,m^{-2}}$
$\mathbf{J}$	Jacobian	
$\mathbf{k}$	Permeability tensor	$\mathrm{m^2}$
k_{rel}	Relative permeability	–
$\mathbf{K}^{(e)}$	Element conductivity matrix	
$\mathbf{L}$	Differential operator	
$\hat{L}$	Approximation operator	
$\mathbf{L}$	Lower matrix	
$L^{(e)}$	Element length	
m	Mass	kg
n	Porosity	$\mathrm{m^3\,m^{-3}}$
$\mathbf{N}^{(e)}$	Element shape function	
Ne	Neumann number, criteria	

Symbol	Parameter	Unit
q^i	Internal heat source	$\mathrm{J\,kg^{-1}\,s^{-1}}$
$\mathbf{q}$	Darcy flux, velocity	$\mathrm{m\,s^{-1}}$
Q	Amount of heat	J
Q_T	Heat production term (volumetric)	$\mathrm{J\,m^{-3}\,s^{-1}}$
q_T	Heat production term (specific)	$\mathrm{kg^{-1}\,s^{-1}}$
$\mathbf{R}$	Residuum vector	
S	Saturation	–
t	Time	s
T	Temperature	K
u	Internal energy	$\mathrm{J\,kg^{-1}}$
$u(t,x)$	Unknown field function of time and space	
u_j^n	Unknown field function approximation at time level n in node j	
$\mathbf{U}$	Upper matrix	
$\mathbf{v}$	Velocity vector	$\mathrm{m\,s^{-1}}$
V	Volume	$\mathrm{m^3}$
$\mathbf{x}$	Solution vector	
Greek symbols		
α	Diffusivity	$\mathrm{m^2\,s^{-1}}$
λ	Thermal conductivity	$\mathrm{W\,K^{-1}\,m^{-1}}$
ρ	Density	$\mathrm{kg\,m^{-3}}$
Δ	Difference	–
ϵ	Volume fraction	–
ϵ_j^n	Approximation error at time level n in node j	–
ψ	Conservation quantity	–
ff	Stress tensor	Pa
μ	Viscosity	Pa s
Exponents, indices		
i,j	Node numbers	
k	Non-linear iteration number	
n	Time level	
s	Solid	
l	Liquid	
w	Water	
f	Fluid	
α	All phases	
γ	Fluid phases	

Appendix B
Keywords

This section provides a wrap-up compendium of the OGS keywords used in this tutorial. A more comprehensive compilation of OGS keywords you can find at www.opengeosys.org/help/documentation/.

B.1 GLI—Geometry

Listing B.1 GLI keyword

```
#POINTS                 // points keyword
0 0 0 0 $NAME POINT0    // point number | x | y | z | point name
1 1 0 0 $NAME POINT1    // point number | x | y | z | point name
#POLYLINE               // polyline keyword
 $NAME                  // polyline name subkeyword
  LINE                  // polyline name
 $POINTS                // polyline points subkeyword
  0                     // point of polyline
  1                     // dito
#STOP                   // end of input data
```

OGS Weblink:

http://www.opengeosys.org/help/documentation/geometry-file

B.2 MSH—Finite Element Mesh

Listing B.2 MSH keyword

```
#FEM_MSH      // file/object keyword
$NODES        // node subkeyword
61            // number of grid nodes
0 0 0 0       // node number x y z
```

© Springer International Publishing Switzerland 2016
N. Böttcher et al., *Geoenergy Modeling I*, SpringerBriefs in Energy,
DOI 10.1007/978-3-319-31335-1

```
1 0  0  1        // dito
...
59       0        0        59
60       0        0        60
$ELEMENTS        // element subkeyword
60               // number of elements
0        0       line    0        1 // element number | material group number |
        element type | element node numbers
1        0       line    1        2 // dito
...
58       0       line    58       59 // dito
59       0       line    59       60 // dito
#STOP            // end of input data
```

OGS Weblink:

   http://www.opengeosys.org/help/documentation/mesh-file

B.3 PCS—Process Definition

Listing B.3 PCS keyword

```
#PROCESS                 // process keyword
  $PCS_TYPE              // process type subkeyword
    HEAT_TRANSPORT       // specified process(es)
    GROUNDWATER_FLOW     // dito
    LIQUID_FLOW          // dito
              ...
#STOP                    // end of input data
```

OGS Weblink:

   www.opengeosys.org/help/documentation/process-file

B.4 NUM—Numerical Properties

Listing B.4 NUM keyword

```
#NUMERICS            // process keyword
  $PCS_TYPE          // process type subkeyword, see PCS above
  $LINEAR_SOLVER     // linear solver type subkeyword, see table below
        Parameters   // 7 parameters, see table below
#STOP                // end of input data
```

Numerical properties

- Linear solver type (its C++ ;-) numbering -1)

  1. SpGAUSS, direct solver
  2. SpBICGSTAB
  3. SpBICG
  4. SpQMRCGSTAB

   5. SpCG
   6. SpCGNR
   7. CGS
   8. SpRichard
   9. SpJOR
  10. SpSOR

- Convergence criterion (its C++ ;-) numbering -1)

   1. Absolutely error $||r|| < \epsilon$
   2. $||r|| < \epsilon ||b||$
   3. $||rn|| < \epsilon ||rn - 1||$
   4. if $||rn|| < 1$ then $||rn|| < \epsilon ||rn - 1||$ else $||r|| < \epsilon$
   5. $||rn|| < \epsilon ||x||$
   6. $||rn|| < \epsilon \max ||rn - 1||, ||x||, ||b||$

- Error tolerance ϵ, according to the convergence criterion model above
- Maximal number of linear solver iterations
- Relaxation parameter $\theta \in [0, 1]$
- Preconditioner

  0 No preconditioner,
  1 Jacobi preconditioner,
  100   ILU preconditioner.

- Storage model

  2 unsymmetrical matrix,
  4 symmetrical matrix.

OGS Weblink:
  http://www.opengeosys.org/help/documentation/numeric-file

B.5 TIM—Time Discretization

Listing B.5 TIM keyword

```
#TIME_STEPPING      // timt stepping keyword
  $PCS_TYPE         // process subkeyword
    HEAT_TRANSPORT  // specified process
  $TIME_STEPS       // time steps subkeyword
    1000 390625e+0  // number of times steps | times step length
  $TIME_END         // end time subkeyword
    1E99            // end time value
  $TIME_START       // starting time subkeyword
    0.0             // starting time value
  $TIME_UNIT        // specified time unit
    DAY             // SECOND, DAY, YEAR
#STOP               // end of input data
```

OGS Weblink:
  http://www.opengeosys.org/help/documentation/time-step-control-file

B.6 IC—Initial Conditions

Listing B.6 IC keyword

```
#INITIAL_CONDITION   // initial conditions keyword
 $PCS_TYPE           // process subkeyword
  HEAT_TRANSPORT     // specified process
 $PRIMARY_VARIABLE   // primary variable subkeyword
  TEMPERATURE1       // specified primary variable
 $GEO_TYPE           // geometry subkeyword
  DOMAIN             // specified geometry: entire domain (all nodes)
 $DIS_TYPE           // distribution subkeyword
  CONSTANT   0       // specified distribution: constant value 0 at DOMAIN
     geometry
#STOP                // end of input data
```

OGS Weblink:

http://www.opengeosys.org/help/documentation/initial-condition-file

B.7 BC—Boundary Conditions

Listing B.7 BC keyword

```
#BOUNDARY_CONDITION // boundary condition keyword
 $PCS_TYPE          // process type subkeyword
  HEAT_TRANSPORT    // specified process
 $PRIMARY_VARIABLE  // primary variable subkeyword
  TEMPERATURE1      // specified primary variable
 $GEO_TYPE          // geometry type subkeyword
  POINT POINT0      // specified geometry type | geometry name
 $DIS_TYPE          // boundary condition type subkeyword
  CONSTANT   1      // boundary condition type | value
#STOP               // end of input data
```

OGS Weblink:

Weblink: http://www.opengeosys.org/help/documentation/boundary-condition-file

B.8 ST—Source/Sink Terms

Listing B.8 ST keyword

```
#SOURCE_TERM               // source term keyword
 $PCS_TYPE                 // process type subkeyword
  LIQUID_FLOW              // specified process
 $PRIMARY_VARIABLE         // primary variable subkeyword
  PRESSURE1                // specified primary variable
 $GEO_TYPE                 // geometry type subkeyword
  POINT POINT0             // specified geometry type | geometry name
 $DIS_TYPE                 // boundary condition type subkeyword
```

Table B.1 Density models

Model	Meaning	Formula	Parameters
0	Curve	RFD file	
1	Constant value	ρ_0	Value of ρ_0
2	Pressure dependent	$\rho(p) = \rho_0(1 + \beta_p(p - p_0))$	ρ_0, β_p, p_0
3	Salinity dependent	$\rho(C) = \rho_0(1 + \beta_C(C - C_0))$	ρ_0, β_p, C_0
4	Temperature dependent	$\rho(p) = \rho_0(1 + \beta_T(T - T_0))$	ρ_0, β_T, T_0
...	...	...	...

```
CONSTANT_NEUMANN 1E-6 // source term type | value
#STOP                 // end of input data
```

OGS Weblink:

http://www.opengeosys.org/help/documentation/source-term-file

B.9 MFP—Fluid Properties

Listing B.9 MFP keyword

```
#FLUID_PROPERTIES            // fluid properties keyword
 $DENSITY                    // fluid density subkeyword
  4 1000 0 -0.2              // type (4: temperature dependent) | 2 values
 $VISCOSITY                  // fluid viscosity subkeyword
  1 0.001                    // type (1: constant value) | value
 $SPECIFIC_HEAT_CAPACITY     // specific heat capacity subkeyword
  1 4200.0                   // type (1: constant value) | value
 $HEAT_CONDUCTIVITY          // thermal heat conductivity subkeyword
  1 0.65                     // type (1: constant value) | value
#STOP                        // end of input data
```

See Table B.1.
OGS Weblink:
http://www.opengeosys.org/help/documentation/fluid-properties-file

See Table B.1.

B.10 MSP—Solid Properties

Listing B.10 MSP keyword

```
#SOLID_PROPERTIES // solid properties keyword
 $DENSITY         // solid density subkeyword
  1 2500          // type (1: constant value) | value
 $THERMAL         // thermal properties subkeyword
  EXPANSION:      // thermal expansion
```

```
    1.0e-5          // values
 CAPACITY:          // heat capacity
    1 1000          // type (1: constant value) | value
 CONDUCTIVITY:      // thermal conductivity
    1 3.2           // type (1: constant value) | value
#STOP               // end of input data
```

OGS Weblink:
  TBD

B.11 MMP—Porous Medium Properties

Listing B.11 MMP keyword

```
#MEDIUM_PROPERTIES      // solid properties keyword
 $GEOMETRY_DIMENSION    // dimension subkeyword
   1                    // 1: one-dimensional problem
 $GEOMETRY_AREA         // geometry area subkeyword
   1.0                  // value in square meter if 1D
 $POROSITY              // porosity subkeyword
   1  0.10              // type (1: constant value) | value
 $STORAGE               // storativity subkeyword
   1  0.0               // type (1: constant value) | value
 $TORTUOSITY            // tortuosity subkeyword
   1  1.000000e+000     // type (1: constant value) | value
 $PERMEABILITY_TENSOR   // pemeability subkeyword
   ISOTROPIC   1.0e-15  // tensor type (ISOTROPIC) | value(s)
 $HEAT_DISPERSION       // porosity subkeyword
   1  0.0e+00 0.0e+00   // type (1: constant values) | longitudinal |
     transverse
                        // thermal dispersion length
#STOP                   // end of input data
```

OGS Weblink:
  http://www.opengeosys.org/help/documentation/material-properties-file

B.12 OUT—Output Parameters

Listing B.12 OUT keyword

```
#OUTPUT           // output keyword
 $PCS_TYPE        // process subkeyword
  HEAT_TRANSPORT  // specified process
 $NOD_VALUES      // nodal values subkeyword
  TEMPERATURE1    // specified nodal values
 $GEO_TYPE        // geometry type subkeyword
  POLYLINE ROCK   // geometry type and name
 $TIM_TYPE        // output times subkeyword
  STEPS 1         // output methods and parameter
#STOP             // end of input data
```

OGS Weblink:
  http://www.opengeosys.org/help/documentation/output-control-file

References

T. Agemar, J. Weber, R. Schulz, Deep geothermal energy production in Germany. Energies 7(7), 4397 (2014). doi:10.3390/en7074397. ISSN:1996–1073

S. Bauer, C. Beyer, F. Dethlefsen, P. Dietrich, R. Duttmann, M. Ebert, V. Feeser, U. Görke, R. Köber, O. Kolditz, W. Rabbel, T. Schanz, D. Schäfer, H. Würdemann, A. Dahmke, Impacts of the use of the geological subsurface for energy storage: an investigation concept. Envron. Earth Sci. 70(8), 3935–3943 (2013)

J. Bear, *Dynamics of Fluids in Porous Media* (Dover Publications, Inc., New York, 1972)

N. Böttcher, G. Blöcher, M. Cacace, O. Kolditz, Heat Transport. In: Kolditz, O. et al. *Thermo-Hydro-Mechanical-Chemical Processes in Fractured Porous Media*. Lecture Notes in Computational Science and Engineering, vol. 86, (Springer, 2012), pp. 89–106

R.M. Bowen, *Theory of Mixture, Continuum Physics*, vol. III, ed. by A.C. Eringen (Academic, New York, 1976)

R.M. Bowen, Incompressible porous media models by use of the theory of mixtures. Int. J. Eng. Sci. 18, 1129–1148 (1980)

M. Cacace, G. Blöcher, N. Watanabe, I. Moeck, N. Börsing, M. Scheck-Wenderoth, O. Kolditz, E. Huenges, Modelling of fractured carbonate reservoirs: outline of a novel technique via a case study from the Molasse Basin, Southern Bavaria, Germany. Environ. Earth Sci. 70(8), 3585–3602 (2013)

H.S. Carslaw, J.C. Jaeger, *Conduction of Heat in Solids*, 2nd edn. (Oxford University Press, Oxford, 1959)

C. Clauser, *Thermal Signatures of Heat Transfer Processes in the Earth's Crust* (Springer, New York, 1999). ISBN:3-540-65604-9

R. de Boer, *Theory of Porous Media* (Springer, Berlin, 2000)

R. de Boer, W. Ehlers, On the problem of fluid- and gas-filled elasto-plastic solids. Int. J. Sol. Struct. 22, 1231–1242 (1986)

M.H. Dickson, M. Fanelli, What is geothermal energy? Technical report, Istituto di Geoscienze e Georisorse, Pisa (2004)

H.-J. Diersch, *FEFLOW - Finite Element Modeling of Flow, Mass and Heat Transport in Porous and Fractured Media* (Springer, Berlin, 2014)

W. Ehlers, J. Bluhm, *Porous Media: Theory, Experiments and Numerical Applications* (Springer, Berlin, 2002)

J.W. Elder, Thermal convection. J. Geol. Soc. 133, 293–309 (1977)

C. Geuzaine, J.-F. Remacle, Gmsh: a 3-d finite element mesh generator with built-in pre- and post-processing facilities. Int. J. Numer. Methods Eng. 79(11), 1309–1331 (2009). doi:10.1002/nme.2579

© Springer International Publishing Switzerland 2016
N. Böttcher et al., *Geoenergy Modeling I*, SpringerBriefs in Energy,
DOI 10.1007/978-3-319-31335-1

U.-J. Goerke, S. Kaiser, A. Bucher, et al. A consistent mixed finite element formulation for hydro-mechanical processes in saturated porous media at large strains based on a generalized material description. Eur. J. Mech. A-Solids **32**, 88–102 (2012)

W. Hackbusch, *Iterative Lösung großer schwach besetzter Gleichungssysteme*. Teubern-Studienbücher: Mathematik, Stuttgart (1991). ISBN:3-519-02372-5

F. Häfner, D. Sames, H.-D. Voigt, *Wärme- und Stofftransport: Mathematische Methoden* (Springer, Berlin, 1992)

R. Helmig, *Multiphase Flow and Transport Processes in the Subsurface* (Springer, Berlin, 1997)

E. Huenges, *Geothermal Energy Systems: Exploration, Development, and Utilization* (Wiley-VCH, Weinheim, 2011). ISBN:978-3-527-64461-2

E. Huenges, T. Kohl, O. Kolditz, J. Bremer, M. Scheck-Wenderoth, T. Vienken, Geothermal energy systems: research perspective for domestic energy provision. Environ. Earth Sci. **70**(8), 3927–3933 (2013)

J. Istok, *Groundwater Modeling by the Finite Element Method*. Water Resources Monograph (American Geophysical Union, 1989). doi:10.1029/WM013

P. Knabner, L. Angermann, *Numerik Partieller Differentialgleichungen: Eine anwendungsorientierte Einführung* (Springer, Berlin/Heidelberg/New York, 2000). ISBN:3-540-66231-6

O. Kolditz, *Strömung, Stoff- und Wärmetransport im Kluftgestein* (Borntraeger-Verlag, Berlin/Stuttgart, 1997)

O. Kolditz, *Computational Methods in Environmental Fluid Mechanics*. Graduate Text Book (Springer Science Publisher, Berlin, 2002)

O. Kolditz, U.-J. Görke, H. Shao, W. Wang, *Thermo-Hydro-Mechanical-Chemical Processes in Fractured Porous Media*. Lecture Notes in Computational Science and Engineering, vol. 86 (Springer, 2012)

O. Kolditz, L.A. Jakobs, E. Huenges, T. Kohl, Geothermal energy: a glimpse at the state of the field and an introduction to the journal. Geotherm. Energy **1** (2013). doi:10.1186/2195-9706-1-1

O. Kolditz, H. Shao, W. Wang, S. Bauer, *Thermo-Hydro-Mechanical-Chemical Processes in Fractured Porous Media: Modelling and Benchmarking - Closed Form Solutions*. Terrestrial Environmental Sciences, vol. 1 (2015a), Springer pp. 1–315

O. Kolditz, H. Xie, Z. Hou, P. Were, H. Zhou, *Subsurface Energy Systems in China: Production, Storage and Conversion*. Environmental Earth Sciences, vol. 73 (2015b), pp. 6727–6732

Y. Kong, Z. Pang, H.B. Shao, O. Kolditz, Recent studies on hydrothermal systems in China: a review. Geothermal Energy **2** (2014). doi:10.1186/s40517-014-0019-8

R.W. Lewis, B.A. Schrefler, *The Finite Element Method in the Static and Dynamic Deformation and Consolidation of Porous Media* (John Wiley 1998)

R.W. Lewis, P. Nithiarasu, K. Seetharamu, *Fundamentals of the Finite Element Method for Heat and Fluid Flow* (John Wiley 2004)

A. Ogata, R.B. Banks, A solution of the differential equation of longitudinal dispersion in porous media. Technical report, U.S. Geological Survey, Washington, DC (1961)

Z. Pang, S. Hu, J. Wang, A roadmap to geothermal energy in China. Sci. Technol. Rev. **30**, 18–24 (2012)

Z. Pang, J. Luo, J. Pang, Towards a new classification scheme of geothermal systems in China, in *Proceedings World Geothermal Congress 2015*, Melbourne (2015)

P. Prevost, Mechanics of continuous porous media. Int. J. Eng. Sci **18**, 787–800 (1980)

M. Scheck-Wenderoth, D. Schmeisser, M. Mutti, O. Kolditz, E. Huenges, H.-M. Schultz, A. Liebscher, M. Bock, Geoenergy: new concepts for utilization of geo-reservoirs as potential energy sources. Environ. Earth Sci. **70**(8), 3427–3431 (2013)

A.E. Scheidegger, General theory of dispersion in porous media. J. Geophys. Res. **66**(10), 3273–3278 (1961). doi:10.1029/JZ066i010p03273. ISSN:0148-0227

H. Schwetlick, H. Kretzschmar, *Numerische Verfahren für Naturwissenschaftler und Ingenieure: Eine computerorientierte Einführung* (Fachbuchverlag Leipzig, Leipzig, 1991)

H. Tenzer, C.-H. Park, O. Kolditz, C.I. McDermott, Application of the geomechanical facies approach and comparison of exploration and evaluation methods used at soultz-sous-forêts (France) and spa urach (Germany) geothermal sites. Environ. Earth Sci. **61**(4), 853–880 (2010)

C. Truesdell, R.A. Toupin, *The Classical Field Theories. Handbuch der Physik*, vol. III/1, ed. by S. Flügge (Springer, Berlin, 1960)

N. Watanabe, W. Wang, C.I. McDermott, T. Taniguchi, O. Kolditz, Uncertainty analysis of thermo-hydro-mechanical coupled processes in heterogeneous porous media. Comput. Mech. **45**(4), 263–280 (2010)

N. Watanabe, W. Wang, J. Taron, U.J. Görke, O. Kolditz, Lower-dimensional interface elements with local enrichment: application to coupled hydro-mechanical problems in discretely fractured porous media. Int. J. Numer. Methods Eng. **90**(8), 1010–1034 (2012)

P. Wriggers, *Nichtlineare Finite-Element-Methoden* (Springer, Berlin/Heidelberg, 2001). doi:10.1007/978-3-642-56865-7. ISBN:978-3-540-67747-5

B. Zehner, N. Watanabe, O. Kolditz, Visualization of gridded scalar data with uncertainty in geosciences. Comput. Geosci. **36**(10), 1268–1275 (2010)

Printed in the United States
By Bookmasters